High Pressure Liquid Chromatography in Clinical Chemistry

Proceedings of a Symposium held at
Kings College Hospital Medical School December 15–16, 1975

High Pressure Liquid Chromatography in Clinical Chemistry

Edited by

P. F. DIXON
C. H. GRAY
C. K. LIM
M. S. STOLL

1976

ACADEMIC PRESS
London   New York   San Francisco

A Subsidiary of Harcourt Brace Jovanovich, Publishers

ACADEMIC PRESS INC. (LONDON) LTD.
24/28 Oval Road,
London NW1

United States Edition published by
ACADEMIC PRESS INC.
111 Fifth Avenue,
New York, New York 10003

Copyright © 1976 by
ACADEMIC PRESS INC. (LONDON) LTD.
Pages 201–209 Crown Copyright

All Rights Reserved
No part of this book may be reproduced in any form by photostat, microfilm, or any other means, without written permission from the publishers

Library of Congress Catalog Card Number: 76-6745
ISBN: 0-12-218450-5

Printed in Great Britain by
J. W. Arrowsmith Ltd., Bristol

CONTRIBUTORS AND PARTICIPANTS

R.F. ADAMS, *The Perkin-Elmer Corporation, Main Avenue, Norwalk, Connecticut, 06856, U.S.A.*

M.W. ANDERS, *Department of Pharmacology and Pediatrics, Division of Clinical Pharmacology, University of Minnesota Health Sciences Centre, 105 Millard Hall, Minneapolis, 55455, U.S.A.*

D.C. AYRES, *Department of Chemistry, Westfield College, University of London, Hampstead, London NW3 7ST, U.K.*

A. AZIZ, *Wolfson Bioanalytical Centre, University of Surrey, Guildford, Surrey GU2 5XH, U.K.*

A.A.-B. BADAWY, *Addiction Unit Research Laboratory, Whitchurch Hospital, Cardiff CF4 7XB, U.K.*

P.B. BANHAM, *I.C.I. Ltd., Central Toxicology Laboratory, Alderley Park, Nr. Macclesfield, Cheshire SK10 4TJ, U.K.*

I.E. BARAKAT, *Department of Chemistry, University College, P.O. Box 78, Cardiff CF1 1XL, U.K.*

S. BARNES, *Department of Medicine, Royal Free Hospital, Pond Street, Hampstead, London NW3, U.K.*

P. BATTEN, *I.C.I. Ltd., Central Toxicology Laboratory, Alderley Park, Nr. Macclesfield, Cheshire SK10 4TJ, U.K.*

A.R. BATTERSBY, *University Chemical Laboratory, Lensfield Road, Cambridge CB2 1EW, U.K.*

T. BAILLIE, *Department of Clinical Pharmacology, Royal Post Graduate Medical School, Hammersmith Hospital, London W12, U.K.*

D.J. BERRY, *Poisons Unit, New Cross Hospital, Avonley Road, London SE14 5ER, U.K.*

Contributors and Participants

D.I. BILKUS, *Glaxo Research Ltd., Sefton Park, Stoke Poges, Buckinghamshire, SL2 4DZ, U.K.*

B.H. BILLING, *Medical Unit, Royal Free Hospital, Pond Street, Hampstead, London NW3 2QG, U.K.*

R. BLAND, *Biochemical Research Department, Allen and Hanburys Research Ltd., Ware, Hertfordshire, U.K.*

H. BRATT, *I.C.I. Ltd., Central Toxicology Laboratory, Alderley Park, Nr. Macclesfield, Cheshire SK10 4TJ, U.K.*

J. BRENDER, *Kobenhavns Kommues Hospital, Copenhagen, Denmark.*

R.R. BRODIE, *Department of Drug Metabolism, Huntingdon Research Centre, Alconbury, Huntingdon, Cambridgeshire, U.K.*

B.N. BROOK, *Whatman Ltd., Springfield Mill, Maidstone, Kent ME14 2LE, U.K.*

D.G. BUCKLEY, *University Chemical Laboratory, Lensfield Road, Cambridge CB2 1EW, U.K.*

R.D. BURNETT, *The Dyson Perrins Laboratory, South Parks Road, Oxford OX1 3QY, U.K.*

R.F. BURNS, *Department of Health and Social Security, 14 Russell Square, London WC1, U.K.*

J. BUTLER, *Department of Chemical Pathology, King's College Hospital, London SE5 8RX, U.K.*

P. CAREY, *Biochemical Research Department, Allen and Hanburys Research Ltd., Ware, Hertfordshire, U.K.*

R.E. CARLSON, *Department of Chemistry, The University of British Columbia, Vancouver, British Columbia V6T 1W5, Canada.*

J.K. CARTER, *District Laboratory, Preston Hall Hospital, Maidstone, Kent ME20 7NH, U.K.*

T. CARTER, *Wolfson Research Laboratories, Queen Elizabeth Medical Centre, Birmingham B15 2TH, U.K.*

J.A. CHRISTOFIDES, *The Epsom Hospital Laboratories, West Park Hospital, Epsom, Surrey, U.K.*

Contributors and Participants

M.J. COOPER, *Department of Pharmacology, University of Minnesota, 105 Millard Hall, Minneapolis, 55455, U.S.A.*

D. COURT, *The City University, London EC1, U.K.*

P.J. COX, *Chester-Beatty Research Institute, Fulham Road, London SW3 6JB, U.K.*

J. CRAYFORD, *Shell Research Ltd., Sittingbourne, Kent, U.K.*

C.J. DAVIS, *Beecham Pharmaceuticals Research Division, Brockham Park, Betchworth, Surrey, U.K.*

D. DELL, *Hoechst Pharmaceutical Research, Walton Manor, Milton Keynes, Buckinghamshire, U.K.*

F. DE MATTEIS, *M.R.C. Toxicology Unit, Woodmansterne Road, Carshalton, Surrey.*

C. DINGWELL, *Sterling-Winthrop Research and Development, Edgefield Avenue, Fawdon, Newcastle-upon-Tyne NE3 3TT, U.K.*

P.F. DIXON, *Department of Biochemistry, Bromley Hospital, Bromley, Kent, U.K.*

L.J. DOLLMAN, *Department of Drug Metabolism, Pfizer Central Research, Sandwich, Kent, U.K.*

D. DOLPHIN, *Department of Chemistry, The University of British Columbia, Vancouver, British Columbia V6T 1N5, Canada.*

P.R. DRAPER, *Reckitt and Colman, Pharmaceutical Division, Pharmaceutical Development Department, Dansom Lane, Hull, U.K.*

G.H. ELDER, *Department of Medical Biochemistry, Welsh National School of Medicine, Heath Park, Cardiff, U.K.*

M.M. ELZAHBY, *Azhar University, Cairo, Egypt.*

J. ENTICKNAP, *Consultant Pathologist, Whipps Cross Hospital, London E11 1NR, U.K.*

R.S. ERSSER, *Department of Chemical Pathology, Institute of Child Health, 30 Guildford Street, London WC1N 1EH, U.K.*

M. EVANS, *Whatman Ltd., Springfield Mill, Maidstone, Kent ME14 2LE, U.K.*

S.E. EVANS, *Clinical Chemistry Department, Queen Elizabeth Hospital, Edgbaston, Birmingham B15 2TH, U.K.*

R. FAIRCHILD, *The Boots Company Ltd., Research Department, Pennyfoot Street, Nottingham NG2 3AA, U.K.*

V. FANTL, *Department of Chemical Pathology, King's College Hospital Medical School, London SE5 8RX, U.K.*

T. FEIZI, *Clinical Research Centre, Harrow, Middlesex, U.K.*

R.J. FLANAGAN, *Poisons Unit, New Cross Hospital, Avonley Road, London SE14 5ER, U.K.*

M. FLINT, *Westwood Hospital, Beverley, Humberside HU17 8BU, U.K.*

J. FRANCIS, *M.R.C. Toxicology Unit, Woodmansterne Road, Carshalton, Surrey, U.K.*

S.B. FRASER, *Unilever Research Laboratories, Port Sunlight, Wirral, Merseyside, U.K.*

D.E.W. FRY, *Epsom Hospital Laboratory, West Park Hospital, Epsom, Surrey, U.K.*

R.G.C. GALLOP, *M.R.E., Porton Down, Salisbury SP4 0JG, U.K.*

M. GAY, *Pharmacokinetics Section, Safety of Medicines Department, I.C.I. Pharmaceuticals Division, Mereside, Alderley Park, Nr. Macclesfield, Cheshire SK10 4TG, U.K.*

J. GILTROW, *Whatman Ltd., Springfield Mill, Maidstone, Kent ME14 2LE, U.K.*

R. GINMAN, *Pharmacy Department, Brighton Polytechnic, Brighton BN2 4GJ, U.K.*

V. GOLDBERG, *6 Hollycroft Avenue, Wembley, Middlesex HA9 8LF, U.K.*

R. GOPALAN, *Department of Chemistry, Westfield College, University of London, Hampstead, London NW3 7ST, U.K.*

A.E.P. GORVIN, *Home Office Central Research Establishment, Aldermaston, Reading, Berkshire, U.K.*

P. GOUGH, *Division of Communicable Diseases, M.R.C. Clinical Research Centre, Watford Road, Harrow, Middlesex, U.K.*

J.L. GOWER, *Department of Chemistry, University College, Cardiff CF1 1XL, U.K.*

Contributors and Participants

B. GRANDCHAMP, *Hospital Louis Mourier, Laboratoire de Biochimie, 178 rue des Renouillers, 92700 Colombes, France.*

C.H. GRAY, *Department of Chemical Pathology, King's College Hospital Medical School, London SE5 8RX, U.K.*

S.P. GRAY, *Department of Biochemistry, R.N. Hospital, Haslar, Gosport, Hampshire PO12 2AA, U.K.*

A. GREEN, *Clinical Chemistry Department, Children's Hospital, Ladywood Middleway, Birmingham 16 8ET, U.K.*

T. GREEN, *I.C.I. Ltd., Central Toxicology Laboratory, Alderley Park, Nr. Macclesfield, Cheshire SK10 4TJ, U.K.*

J.M.C. GUTTERIDGE, *Department of Clinical Chemistry, Whittington Hospital, Archway Road, London N19, U.K.*

M.R. HARDEMAN, *Academic Hospital "Wilhelmina Gasthuis", Department Intern. Medicine, Clinical Chemistry Division, 1e Helmersstraat 104, Amsterdam, The Netherlands.*

F.E. HARPER, *Department of Clinical Pathology, General Hospital, Middlesbrough, Cleveland TS5 5AZ, U.K.*

M. HARRIS, *Queen Elizabeth Hospital for Children, Biochemistry Department, Hackney Road, London E2 8PS, U.K.*

A.D.R. HARRISON, *Wolfson Bioanalytical Centre, University of Surrey, Guildford, Surrey GU2 5XH, U.K.*

H. HEID, *Forschungsinstitut für experimentelle Ernährung e.V., 8520 Erlangen, Langermarckplatz 5½*

J.T. HINDMARSH, *Pathological Institute, University Avenue, Halifax, Nova Scotia, Canada.*

K. HOBBS, *The Radiochemical Centre, Amersham, Buckinghamshire, U.K.*

G.L. HODGSON, *University Chemical Laboratory, Lensfield Road, Cambridge CB2 1EW. U.K.*

M. HOSSAIN, *Chemistry Department, Westfield College, University of London, Hampstead, London NW3 7ST, U.K.*

G.W. HOUGHTON, *Metabolism Department, Pharmaceutical Research, May and Baker Ltd., Dagenham, Essex, U.K.*

I.M. HOUSE, *Poisons Unit, New Cross Hospital, Avonley Road, London SE14, U.K.*

J.F.K. HUBER, *Laboratory for Analytical Chemistry, University of Amsterdam, The Netherlands.*

P. HUBERT, *Simon Stevin Instituut, Jerusalemstraat 34, B-8000, Brugge, Belgium.*

J.T. IRELAND, *Biochemistry Department, Alder Hey Children's Hospital, Liverpool L12 2AP, U.K.*

A.H. JACKSON, *Department of Chemistry, University College, Cardiff CF1 1XL, U.K.*

B. JOHNS, *Department of Drug Metabolism, Pfizer Ltd., Sandwich, Kent, U.K.*

G.F. JOHNSTON, *Waters Associates (Instruments) Ltd., Vauxhall Works, Greg Street, South Reddish, Stockport, Cheshire SK5 7BR, U.K.*

C. JOLLEY, *Waters Associates (Instruments) Ltd., Vauxhall Works, Greg Street, South Reddish, Stockport, Cheshire SK5 7BR, U.K.*

K.M. JONES, *Chemical Pathology, Royal Alexandra Hospital, Rhyl, Clwyd, North Wales, U.K.*

K.P. JONES, *Department of Photobiology, Institute of Dermatology, Homerton Grove, London C08 5BA, U.K.*

J. JURAND, *Wolfson Liquid Chromatography Unit, Department of Chemistry, University of Edinburgh, West Mains Road, Edinburgh EH9 3JJ, U.K.*

D.N. KIRK, *Chemistry Department, Westfield College, University of London, Hampstead, London NW3 7ST, U.K.*

W.E. KLINGE, *Department of Biochemistry, Charing Cross Hospital Medical School, London W6 8RF, U.K.*

D.W. KNIGHT, *Chemistry Department, University College, Cardiff CF1 1XL, U.K.*

J.H. KNOX, *Wolfson Liquid Chromatography Unit, Department of Chemistry, University of Edinburgh, West Mains Road, Edinburgh EH9 3JJ, U.K.*

R. KOENIGSBERGER, *15 The Mount Square, London NW3 6SX, U.K.*

L.J. KRICKA, *Wolfson Research Laboratories, Department of Clinical Chemistry, Queen Elizabeth Medical Centre, Edgbaston, Birmingham B15 2TH, U.K.*

W.J. LEAHEY, *The Queen's University of Belfast, Department of Therapeutics and Pharmacology, Institute of Clinical Science, Grosvenor Road, Belfast BT12 6BJ, Northern Ireland, U.K.*

T.E.B. LEAKEY, *Queen Elizabeth Hospital for Children, Hackney Road, London, E2, U.K.*

J.P. LEPPARD, *Wolfson Bioanalytical Centre, University of Surrey, Guildford GU2 5XH, Surrey, U.K.*

K.O. LEWIS, *Clinical Chemistry, Dudley Road Hospital, Dudley Road, Birmingham B18 7QH, U.K.*

C.K. LIM, *Department of Chemical Pathology, King's College Hospital Medical School, London SE5 8RX, U.K.*

E.B.C. LLAMBIAS, *University of Buenos Aires, Argentina.*

E. LOCKEY, *Department of Pathology, Barnet General Hospital, Wellhouse Lane, Barnet, Hertfordshire, U.K.*

W.H. LOCKWOOD, *129 High Street, North Sydney, New South Wales, 2060, Australia.*

G.A. LUYTENS, *Packard-Becker B.V., Delft, The Netherlands.*

A. MCBURNEY, *Purine Laboratory, 4th Floor Hunts House, Guy's Hospital Medical School, London SE1 9RT, U.K.*

E. MCDONALD, *University Chemical Laboratory, Lensfield Road, Cambridge CB2 1EW, U.K.*

P.H. MCNALLY, *Reckitt and Colman, Pharmaceutical Division, Chemical Services Department, Dansom Lane, Hull, U.K.*

H.N. MAGNANI, *Department of Experimental Pathology, The Medical School, Birmingham B15 2TU, U.K.*

I.A. MAGNUS, *Department of Photobiology, Institute of Dermatology, Homerton Grove, London CO8 5BA, U.K.*

R. MALEY, *Coulter Electronics Ltd., Bilton Way, Dallow Road, Luton, Bedfordshire.*

R.E. MARKWELL, *University Chemical Laboratory, Lensfield Road, Cambridge CB2 1EW, U.K.*

A.M. MARSDEN, *Clinical Pathology Unit, Safety of Medicines Department, I.C.I. Pharmaceuticals Division, Mereside, Alderley Park, Macclesfield, Cheshire SK10 4TG, U.K.*

A.C. MARSHALL, *Physical and Analytical Unit, Beecham Pharmaceuticals Research Division, Brockham Park, Betchworth, Surrey RH3 7AJ, U.K.*

S.A. MATLIN, *Chemistry Department, The City University, London EC1, U.K.*

D. MEAKIN, *Waters Associates (Instruments) Ltd., Vauxhall Works, Greg Street, South Reddish, Stockport, Cheshire SK5 7BR, U.K.*

J. MEAKIN, *Whatman Ltd., Springfield Mill, Maidstone, Kent ME14 2LE, U.K.*

T.J.X. MEE, *Postgraduate School of Studies in Pharmaceutical Chemistry, University of Bradford, Bradford BD7 1DP, U.K.*

I.S. MENZIES, *Department of Clinical Chemistry, St. Thomas's Hospital, London SE1 7EH, U.K.*

U.A. MEYER, *CL-25 Kantonsspital, Division of Clinical Pharmacology, CH 8091 Zurich, Switzerland.*

B.L. MIRKIN, *Departments of Pharmacology and Pediatrics, Division of Clinical Pharmacology, University of Minnesota Health Sciences Center, 105 Millard Hall, Minneapolis, 55455, U.S.A.*

F.L. MITCHELL, *Clinical Research Centre, Watford Road, Harrow, Middlesex, U.K.*

A.C. MOFFAT, *Home Office Central Research Establishment, Aldermaston, Reading, Berkshire, U.K.*

S.M. MOHAMEDALLY, *Medical Unit, King's College Hospital Medical School, London SE5 8RX, U.K.*

A. MORHAF FATHY, *Azhar University, Cairo, Egypt.*

S.J. MOULTON, *Pye Unicam Ltd., York Street, Cambridge CB1 2PX, U.K.*

J.N. MOUNT, *Clinical Chemistry Department, St. Thomas' Hospital, Lambeth Palace Road, London SE1, U.K.*

Contributors and Participants

A. MUIR, *Biochemistry Department, Royal Prince Alfred Hospital, Missenden Road, Camperdown 2050, New South Wales, Australia.*

D. MURPHY, *Department of Chemical Pathology, Bristol Royal Infirmary, Bristol BS2 8HW, U.K.*

D. MYHILL, *Brocades Ltd., Byfleet, Surrey, U.K.*

G.E. NEAL, *M.R.C. Toxicology Unit, Woodmansterne Road, Carshalton, Surrey, U.K.*

D.C. NICHOLSON, *Department of Chemical Pathology, King's College Hospital Medical School, London SE5 8RX, U.K.*

J. NISBETT, *Department of Chemical Pathology, St. George's Hospital, London SW17, U.K.*

Y. NORDMAN, *Hopital Louis Mourier, Laboratoire de Biochimie, 178 rue des Renouillers, 92700 Colombes, France.*

R.F. NUNN, *Department of Chemical Pathology, Royal Sussex County Hospital, Brighton, Sussex, U.K.*

M.D.G. OATES, *The Group Laboratory, The Royal Infirmary, Wigan, Lancashire, U.K.*

E.M. ODAM, *Pest Infestation Control Laboratory, Ministry of Agriculture, Fisheries and Food, Hook Rise South, Tolworth, Surbiton, Surrey KT6 7NF, U.K.*

P. O'GORMAN, *Brook Hospital, Shooters Hill, London SE18, U.K.*

V. OHANIAN, *Pathology Department, Central Research, Pfizer Ltd., Sandwich, Kent, U.K.*

A.U. PARMAN, *Department of Chemical Pathology, King's College Hospital Medical School, London SE5 8RX, U.K.*

E.S. PARRY, *Leishman Laboratory, Cambridge Military Hospital, Aldershot, Hampshire, U.K.*

J.W. PAXTON, *Department of Materia Medica, Stobhill General Hospital, Glasgow G21 3UW, U.K.*

R.H. PAYNE, *L.K.B. Instruments Ltd., 232 Addington Road, South Croydon, Surrey CR2 8YD, U.K.*

M.A. PEAT, *Department of Forensic Medicine, Charing Cross Hospital Medical School, Fulham Palace Road, London W6 8RF, U.K.*

D. PERRETT, *Medical Unit, St. Bartholomew's Hospital, London EC1A 7BE, U.K.*

Z.J. PETRYKA, *Northwestern Hospital, Chicago Avenue at 27th Street, Minneapolis, Minnesota, 55407, U.S.A.*

G.R. PHILPOT, *Department of Pathology, Frenchay Hospital, Bristol BS16 1LE, U.K.*

P.F. PLUMLEY, *Royal East Sussex Hospital, Hastings, Sussex, U.K.*

J. PLUMRIDGE, *Perkin-Elmer Ltd., Post Office Lane, Beaconsfield, Buckinghamshire HP9 1QA, U.K.*

H. POPPE, *Laboratory for Analytical Chemistry, Nieuwe Achtergracht 166, Amsterdam, The Netherlands.*

D.E. PRYDE, *Clinical Chemistry Department, Northwick Park Hospital, Watford Road, Harrow, Middlesex HA5 2DN, U.K.*

C. QUINN, *Department of Pathology, Mater Misericordia Hospital, Dublin 7, Eire.*

K.R.N. RAO, *Department of Chemistry, University College, P.O. Box 78, Cardiff CF1 1XL, U.K.*

P.J. REEDS, *Department of Chemical Pathology, King's College Hospital Medical School, London SE5 8RX, U.K.*

E. REID, *Wolfson Bioanalytical Centre, University of Surrey, Guildford, Surrey GU2 5XH, U.K.*

P.J. RIDGEON, *Pye Unicam Ltd., York Street, Cambridge CB1 2PX, U.K.*

D.M. RUTHERFORD, *Poisons Unit, New Cross Hospital, Avonley Road, London SE14, U.K.*

S. SHAW, *James A. Jobling and Co. Ltd., Laboratory Division, Stone, Staffordshire, U.K.*

M.J. SHEARER, *Haematology Department, Guy's Hospital, London SE1 9RT, U.K.*

P.J. SIMONS, *Drug Metabolism Section, Safety of Medicines Department, I.C.I. Pharmaceuticals Division, Mereside, Alderley Park, Macclesfield, Cheshire SK10 4TG, U.K.*

R. SKINNER, *Royal Free Hospital, Pond Street, London NW3 2QG, U.K.*

Contributors and Participants xv

A.R. SINAIKO, *Departments of Pharmacology and Pediatrics, Division of Clinical Pharmacology, University of Minnesota Health Sciences Center, 105 Millard Hall, Minneapolis, 55455, U.S.A.*

A. SMITH, *Department of Chemical Pathology, King's College Hospital Medical School, London SE5 8RX, U.K.*

J.A. SMITH, *Department of Pharmaceutical Chemistry, School of Pharmacy, University of Bradford, Bradford, Yorkshire, U.K.*

S.G. SMITH, *Department of Medicine, Welsh National School of Medicine, Heath Park, Cardiff, U.K.*

S.P. SMYTH, *V.A. Howe and Co. Ltd., 88 Peterborough Road, London SW6, U.K.*

D.G. SNELGROVE, *B.P. Research Centre, Middlesex, U.K.*

J. SPAANS, *Packard-Becker B.V., Delft, The Netherlands.*

F.G. STANFORD, *Pharmaceuticals Production Department, The Radiochemical Centre, Amersham, Buckinghamshire, U.K.*

J.F. STEVENS, *Courtauld Institute, Middlesex Hospital Medical School, Riding House Street, London W1, U.K.*

D. STEVENSON, *Shell Research Ltd., Tunstall Laboratory, Broad Oak Road, Sittingbourne, Kent, U.K.*

M.S. STOLL, *Department of Biochemistry, Bromley Hospital, Bromley, Kent, U.K.*

A.T. SULLIVAN, *Home Office Central Research Establishment, Aldermaston, Reading, Berkshire, U.K.*

J.S. TATE, *Beecham Pharmaceuticals, Randals Road, Leatherhead, Surrey, U.K.*

G.G. THOMPSON, *Department of Materia Medica, Stobhill General Hospital, Glasgow G21 3UW, U.K.*

C.J. THRELFALL, *Toxicology Research Unit, M.R.C. Laboratories, Woodmansterne Road, Carshalton, Surrey, U.K.*

J.A. TOVEY, *14 Cottrell Road, Roath, Cardiff CF2 3EY, U.K.*

J.E. TOVEY, *Worthing Hospital, Worthing, Sussex, U.K.*

R. TOWILL, *Department of Chemistry, University College, P.O. Box 78, Cardiff CF1 1XL, U.K.*

B.M. TRACEY, *Department of Chemistry, Westfield College, University of London, Hampstead, London NW3 7ST, U.K.*

P. TREHERNE, *Perkin-Elmer Ltd., Post Office Lane, Beaconsfield, Buckinghamshire HP9 1QA, U.K.*

P. TULLETT, *Searle Scientific Services, Lane End Road, High Wycombe, Buckinghamshire, U.K.*

P.J. TWITCHETT, *Home Office Central Research Establishment, Aldermaston, Reading, Berkshire, U.K.*

F.L. VANDEMARK, *The Perkin-Elmer Corporation, Main Avenue, Norwalk, Connecticutt, 06856, U.S.A.*

G. VAN DER WAL, *Elisabeth's Gasthuis, Boerhaavelaan, Haarlem, The Netherlands.*

SJ. VAN DER WAL, *Laboratory voor Analytical Scheikunde, Nwe Achtergracht 166, Amsterdam, The Netherlands.*

R.S. WARD, *Department of Chemistry, University College, Singleton Park, Swansea SA2 8PP, U.K.*

R. WATTS, *The Medical School, Rheumatism Research Wing, Birmingham B15 2TJ, U.K.*

V. WEAVER, *Whatman Ltd., Springfield Mill, Maidstone, Kent ME14 2LE, U.K.*

B.B. WHEALS, *The Metropolitan Police Forensic Science Laboratory, 109 Lambeth Road, London SE1, U.K.*

P.J. WHITE, *Royal East Sussex Hospital, Cambridge Road, Hastings, Sussex, U.K.*

P.L. WILLIAMS, *Home Office Central Research Establishment, Aldermaston, Reading, Berkshire, U.K.*

P. WILLIAMSON, *Waters Associates (Instruments) Ltd., Vauxhall Works, Greg Street, South Reddish, Stockport, Cheshire SK5 7BR, U.K.*

F.W. WILLMOTT, *Mullard Research Laboratories, Cross Oak Lane, Redhill, Surrey RH1 5HA, U.K.*

T.K. WITH, *St. Jørgensvej 81, 5700 Svendborg, Denmark.*

R.J. VON WITT, *Department of Clinical Chemistry, Guy's Hospital Medical School, London SE1 9RT, U.K.*

D.C.F. WOOD, *Department of Pathology, St. Bernard's Hospital, Southall, Middlesex, U.K.*

N.J. WOODHOUSE, *Medical Unit, King's College Hospital Medical School, London SE5 8RX, U.K.*

R.N. WOODHOUSE, *Department of Drug Metabolism, Huntingdon Research Centre, Alconbury, Huntingdon, Cambridgeshire, U.K.*

J.W. WOOLLEN, *Department of Biochemistry, Edgware General Hospital, Edgware, Middlesex, U.K.*

H.G.J. WORTH, *Department of Clinical Chemistry, Queen Elizabeth Medical Centre, Birmingham B15 2TH, U.K.*

PREFACE

Just over a decade ago, gas liquid chromatography began to be used in clinical chemistry in laboratories. With its ability to combine chromatographic separation with detection and quantitation, its resolution, its sensitivity and its reproducibility, it promised to be one of the most powerful tools for clinical investigation. However, at present, with the exception of a few specialised fields, the gas chromatograph is of secondary importance in most clinical laboratories. The re-birth of column chromatography with pressurised and metered solvent delivery systems to give speed and reproducibility with dense packing materials to permit resolution and with integral detection and quantifying systems, i.e. high performance, high speed or high pressure liquid chromatography (HPLC*) now gives rise to fresh hopes that a new era of biochemical analysis is about to dawn.

HPLC has certain inherent advantages over GLC. In theory, any system which can be applied to paper or thin layer chromatography can be applied also to HPLC. Low temperature working avoids the possibility of thermal decomposition, a high solubility in the mobile phase eliminates the necessity of achieving volatility in a gas phase. Detection and quantitation can be achieved by conventional chemical methods and in many applications these are non-destructive so that the complete fractions can be collected and submitted to further analysis.

In organic chemistry the technique is used to monitor synthetic sequences, to separate and identify reactants and products, which help the understanding of the mechanism and kinetics of a reaction. Pure compounds may be isolated for characterization by physico-chemical methods such as mass spectrometry, IR, UV and NMR spectroscopy.

*One contributor, Dr. Eric Reid and his collaborators, raised the question of terminology. He rightly regarded HPLC as the least objectionable term where "P" could denote "Pressure" or if preferred "Performance", but pleaded for the demise of HELC, HRLC and HSLC, as well as LC alone.

In biochemistry HPLC is used to study biosynthesis of natural products and to follow metabolic pathways. Compounds of biochemical significances are purified for assay by radioimmunoassay, competitive protein binding and other established assay methods.

HPLC is valuable for studying drug metabolism, by following the drug metabolites or by monitoring the unchanged drugs in physiological fluids.

In environmental science HPLC has become an important tool for detecting pollutants in soil, air and water.

This led one of us (P. F. Dixon) to believe that the time was opportune to arrange an informal meeting of clinical biochemists to ascertain what had been done in this field and to consider the direction in which further developments should be made. A preliminary enquiry revealed that such a meeting might attract 30 or more participants and encouraged us to go ahead with arrangements for such a meeting. In the event it became evident that the proposal fulfilled a considerable need, for the attendance reached 180 persons, some of whom came from far and wide as will be seen from the list of participants.

The number and quality of the papers presented showed that HPLC had advantages in the speed with which the apparatus could be set up, the brief equilibration time, its efficacy in separating specific substances from very complex biological fluids and its ability to separate isomeric substances of very similar chemical structure. Some of the papers emphasised the possibility of using inexpensive apparatus, although clearly many of the commercial instruments had much to commend them. One area in which GLC retained the lead was in detectors which in HPLC usually consisted of a variable wavelength flow-through spectrophotometer, although some commercial instruments still made use of the refractometer, and only limited wavelength spectrophotometers. Only one paper referred to the fluorimeter as a detector and it was evident that the sensitivity of the flame ionisation and electron capture detectors, and of the mass spectrometer for GLC, meant that this well-established technique would continue to permit important analysis of substances which can be made volatile. Similarly, the ready linkage to a moving wire detector described by Dr. Worth indicated one direction in which on-line mass spectrometry might be expected to mature.

As HPLC begins to pass from the domain of pure chemistry to that of clinical chemistry it is difficult not to fall into the trap of over-enthusiasm with wild hopes for its future possibilities, but it is also difficult not to believe that 10 years hence HPLC will be much more a part of routine clinical biochemistry than is GLC at present.

P. F. DIXON

Department of Biochemistry, Bromley Hospital,
Bromley, Kent, England

C. H. GRAY

Department of Chemical Pathology,
Kings College Hospital Medical School,
London, England

C. K. LIM

Department of Chemical Pathology,
Kings College Hospital Medical School,
London, England

M. S. STOLL

Department of Biochemistry, Bromley Hospital,
Bromley, Kent, England

ACKNOWLEDGEMENTS

We thank Waters Associates for their generous financial support, Miss Patricia Thornton for taking over the heavy organisation the symposium demanded, Mrs. Rosalind Carpenter and Mrs. Gillian Gaydon for much clerical assistance, and Mrs. Dorothy Blewer and Miss Marie Mitten for looking after the registration and Professors Knox and Jackson for taking the Chair at various parts of the symposium. Miss Joan Butler gave much help during the symposium and in checking proofs and in preparation of the index.

Our thanks are specially due to the Normanby College of Nursing and its Bursar, Mrs. Linda Bashford, for allowing us the use of the splendid lecture theatre and to Academic Press for publishing the proceedings so promptly, especially important with such a fast developing subject.

CONTENTS

Contributors and Participants — v
Preface   *P.F. Dixon, C.H. Gray, C.K. Lim and M.S. Stoll* — xix
Acknowledgements — xxii

High Pressure Liquid Chromatography - The Present Situation   *John H. Knox* — 1

Some Early Experience with a Moving Wire Liquid Chromatograph   *H.G.J. Worth* — 11

Trace Enrichment in Liquid Chromatographic Analysis   *P. Williamson* — 17

The Use of Soft Packing Materials of Small Particle Size in the Separation of Basic Proteins by HPLC   *Sj. van der Wal and J.F.K. Huber* — 21

Rapid Ion-Exchange Chromatography of Plasma Amino-acids   *R.S. Ersser* — 23

Analysis of Serum Lipids and Lipoproteins by High Speed Liquid Chromatography   *T. Carter, H.N. Magnani and Rodney Watts* — 33

Separation of Oligosaccharides   *Ten Feizi, P. Gough and P. Williamson* — 41

Separations of Some Steroids, Triterpenes, Glycosides, Bitter Principles, and Related Compounds   *R.D. Burnett* — 45

Separation of Oestrogens and Determination of Oestriol in Human Pregnancy Urine   *Vera Fantl, C.K. Lim and C.H. Gray* — 51

Separation of Corticosteroids for Measurement by Competitive Protein Binding   *Joan Butler, Vera Fantl and C.K. Lim* — 59

Contents

Separation of Porphyrin Isomers by HPLC – Biochemical and Biosynthetic Applications *Alan R. Battersby, Dennis G. Buckley, Gordon L. Hodgson, Roger E. Markwell and Edward McDonald* 63

HPLC Analysis of Porphyrins in Biological Materials *N. Evans, A.H. Jackson, S.A. Matlin and R. Towill* 71

Porphyrin Profiles in Porphyrias *C.H. Gray, C.K. Lim and D.C. Nicholson* 79

Application of HPLC to the Analysis of Clinically Important Porphyrins *R.E. Carlson and D. Dolphin* 87

The Separation of Bile Pigments by HPLC *M.S. Stoll, C.K. Lim and C.H. Gray* 97

High Performance Liquid Chromatography of Nucleotides in Biological Fluids *D. Perrett* 109

High Pressure Chromatographic Separation of Some Biogenic Amines and Derivatives Implicated in the Aetiology of Schizophrenia *T.J.X. Mee and J.A. Smith* 119

Separation of Catecholamines and their Metabolites by High Speed Liquid Chromatography *J. Jurand* 125

Problems in Separating Urinary Metadrenalines *J.P. Leppard, A.D.R. Harrison and E. Reid* 131

Determination of Anticonvulsants in Serum by Use of High Pressure Liquid Chromatography *Reginald F. Adams and Frank L. Vandemark* 143

Analysis of Carbamazepine in Plasma by High Pressure Liquid Chromatography *I.M. House and D.J. Berry* 155

Drug Levels in Plasma Simultaneous Determination of Phenobarbital and Diphenylhydantoin *G.F. Johnston* 163

The HPLC Detection of Some Drugs Taken in Overdose *P.F. Dixon and M.S. Stoll* 165

Application of HPLC to the Study of the Disposition of Hydrochlorothiazide in Adults and Children *M.J. Cooper, M.W. Anders, A.R. Sinaiko and B.L. Mirkin* 175

Contents   xxv

An Attempt to Isolate and Purify Drug Metabolites   183
  by HPLC for Characterisation by Physico-Chemical
  Methods   *P.J. Simons*

The Determination of the Semi-Synthetic Cephalosporin   185
  HR 580 in Plasma and Urine   *D. Dell*

HPLC Analysis of Chlorophenolic Pollutants and of   195
  their Oxidation Products   *D.C. Ayres and R. Gopalan*

An Evaluation of Some HPLC Columns for the Identifi-   201
  cation and Quantitation of Drugs and Metabolites
  *P.J. Twitchett, A.E.P. Gorvin, A.C. Moffat,
  P.L. Williams and A.T. Sullivan*

Separation of Drugs by HPLC and the Application of   211
  Fluorimetric Detection to Drug Problems
  B.B. Wheals

Appendix   218

Index   221

HIGH PRESSURE LIQUID CHROMATOGRAPHY - THE PRESENT SITUATION

John H. Knox

*Wolfson Liquid Chromatography Unit,
Department of Chemistry, University of Edinburgh.*

Looking back at column liquid chromatography in the 1960's it becomes clear that the main advances which have occurred since that period can be attributed to the development of equipment, packing materials of very high intrinsic performance, and a greatly improved understanding of the chromatographic process.

Equipment

In the late 1960's Huber (1) and Kirkland (2) independently constructed high pressure LC systems which became the basis of commercial instruments. Since then the gross specification of HPLC's has not greatly altered (3).

Kirkland's first photometer used 254 nm radiation (the mercury resonance line) and had a noise level corresponding to about 10^{-4} absorbance units (equivalent to 0.023% transmitted light intensity). The best monochromatic photometers now available are only 2 to 5 times more stable and the problem of drift due mainly to thermal effects has still not been fully solved in commercial instruments. Variable wavelength detectors now common are somewhat less stable than the monochromatic photometers (4). Electrochemical detectors are known to be as sensitive as UV monitors (5) but are still little used.

Pumps, in spite of several years of progress, still pose major problems. Thus the so-called constant flow pumps cannot fully compensate for liquid compressibility over wide ranges of flow rate (6), especially in gradient elution applications, while constant pressure pumps, which are cheaper and more straightforward mechanically, do not give a constant flow. Manufacturers still have to design pumps which have the thousandfold range of flow rate necessary for gradient elution, and at the same time give the preset flow irrespective of the back pressure against which they operate.

Injection is another area where simple yet satisfactory devices are lacking. The standard septum injector, while relatively inexpensive, is unsatisfactory for a number of reasons: particles of septum dislodged by the syringe cause peak broadening and distortion; septa must be frequently replaced and cannot easily be used above 1500 p.s.i.; needles easily become blocked with detached particles of septum and this can then cause glass syringes to split when injection is attempted. Double seal devices and stop flow injectors allow injection by syringe and avoid these problems but they are often cumbersome and require too many individual operations to inject a sample. Injection valves also circumvent the disadvantages of septum injection but they do not generally allow on-column injection.

Devices for sample injection clearly require improvement. Ideally we would like to inject a small sample of 1-10 µl centrally onto the top of the column by syringe or injection valve without involving any elastomer seals or septa.

Some commercial instruments include column thermostats (3), but in others the column is unprotected and mounted on the side of the instrument. Although retention volumes depend on temperature (7) the dependence is not great and ambient temperature operation is satisfactory for most forms of LC. Nevertheless, it is important that the column temperature should be constant from day to day and that thermal gradients within the detector cell should be minimized. Fixed temperature thermostatting at $\sim 30^\circ C$ therefore seems to be desirable.

Column fittings are another area where the present technology based on Swagelok connectors leaves something to be desired. Quick changing of columns is difficult and any injector head can only accommodate a column of a particular diameter, generally $\frac{1}{4}$ in. Then at the column outlet it is generally necessary to use modified Swagelok connectors in order to reduce dead volumes and provide properly tailored flow geometry. Different manufacturers use different modifications. Evidently column fittings specially designed for HPLC are now required to accommodate high performance columns, high quality injectors and columns of a range of diameters.

Column Performance and Operating Parameters

Martin and Synge (8) pointed out in 1941 that substantial improvement in performance of LC would be achieved by using high pressures across the column and with very small particles of packing. This was not fully appreciated until Kirkland pioneered high pressure systems in 1967-1969 by using 30 μm pellicular materials which he had developed specially for high performance LC. Since then smaller and smaller particles have been used and 5-10 μm porous particles have become the preferred materials (9-12). While 30 μm particles were dry-packed, the modern 5 and 10 μm particles are packed by the "slurry method" by driving a slurry of the packing into the column under high pressure (13).

Column efficiency is measured in terms of the number of theoretical plates to which the column is equivalent. If w is the width of the peak at its base and t_R its retention time, the number of plates to which the column is equivalent is

$$N = 16(t_R/w)^2$$

For a column length L, the HETP is

$$H = L/N = (L/16)(w/t_R)^2$$

It can be shown that the best measure of the intrinsic performance of a column packing is not the absolute plate height, H, but the plate height measured in terms of the particle diameter, d_p, that is the reduced plate height

$$h = H/d_p = (1/16)(L/d_p)(w/t_R)^2$$

A good column will show a plate height of not more than 3 particle diameters (12). Unfortunately h depends upon eluent velocity and the minimum h occurs at a specific velocity which is lower the larger d_p. With 5 μm particles one normally works at or near the minimum h but with 30 μm particles the minimum of h occurs at an impracticably low velocity and one is forced to operate in the unfavourable region where $h \sim 10$. Smaller particles therefore give higher speed and better resolution. Typically with 5 or 10 μm particles columns are 100-250 mm in length and give N in the range 3000 to 15,000. Pressure drops for reasonable analysis times range from say 100 to 2000 p.s.i. Since higher plate numbers are rarely required we conclude that the major problems in column dynamics have been solved.

The column bore which is desirable for small particle packings is 5-8 mm. This is much wider than the bore of 2 mm originally employed by Kirkland for 30 μm packings (2). Wide bore columns are employed in order to eliminate wall effects for it can be shown by theory and experiment that samples injected centrally into sufficiently wide columns fail to reach the walls before emerging from the column (14).

Chromatographic Partitioning Systems and Applications

Until the advent of high performance LC the main techniques of LC were adsorption chromatography performed in wide columns or on TLC plates, ion exchange chromatography carried out in columns packed with ion-exchange resins, liquid-liquid partition chromatography carried out on cellulose either in packed columns or as paper chromatography, and exclusion chromatography carried out on porous polymers. These techniques were generally slow and, with the exception of ion exchange chromatography, gave poor resolution.

It is now appreciated that the essential requirements for rapid and efficient chromatography are firstly rapid equilibration of solutes between the mobile zone (outside the particles of column packing) and the static zone (within the particles) and secondly, a distribution coefficient between the zones which for any given solute is independent of solute concentration, and therefore constant throughout the chromatographic band.

The first requirement demands the use of high pressures and small (5-10 μm) rigid particles which rules out packings such as cellulose, soft polymeric gels and to a lesser extent ion-exchange resin beads. The main problem with resin beads, in fact, is not that they are soft but that they can swell in eluents and that mass transfer by diffusion within resin beads is slow. The second requirement is difficult to relate to particle structure and a major problem in HPLC is to achieve highly uniform distribution coefficients over wide concentration ranges for a wide variety of solutes.

Since many of the classical materials cannot be used under high pressures we have had to devise novel partitioning systems and from this has grown a better understanding of the partitioning mechanism. It is now clear, for example, that there is no clear distinction between adsorption and partition chromatography. For example, a common stationary phase in partition chromatography is $\beta\beta'$-oxydiproprionitrile (BOP) which is held by a porous support and employed with hexane as eluent. Such two-phase columns are unstable because of solution of the stationary in the mobile phase. It turns out that almost identical separations can be achieved by a much simpler technique whereby BOP saturated hexane is fed to a column containing silica gel. The silica gel (which may have a pore diameter between 50 and 100 Å) adsorbs the BOP to give a layer of BOP at the surface. The main role of the silica gel is not to adsorb the solute itself but to adsorb a component from the eluent. Such chromatographic systems are both stable and reproducible, unlike the older two phase systems.

It seems likely that in adsorption chromatography with

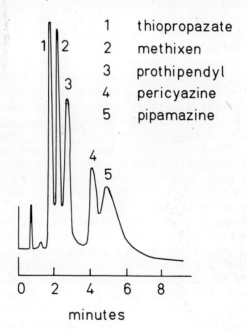

Fig. 1. Separation of phenothiazines by adsorption chromatography. Adsorbent 20 μm Spherisorb AY; Column 125 x 5 mm; eluent methylene chloride-pentane-acetic acid 68:25:7 v/v; detection UV photometer at 254 nm. Solute identification: (1) thiopropazate, (2) methixen, (3) prothipendyl, (4) pericyazine, (5) pipamazine.
(Reproduced by permission of J. Chromatog. from Ref. 15).

mixed eluents the same situation exists where separation is largely due to partition of solutes between the eluent and a surface phase which is rich in the more polar components of the eluent.

The use of an adsorbent in the partition mode is now a standard method of chromatography and its use for the separation of tranquillizers (15) is shown in Figure 1, but there are certain problems when trying to separate complex polar molecules. While the early peaks in Figure 1 possess excellent symmetry the final peak is badly tailed. This is indicative of a poor distribution isotherm caused by specific adsorption of the solute onto fortuitously favourable sites on the silica surface. Such tailing nearly always occurs with well retained solutes whose molecules contain several polar functional groups such as $-NH_2$, $-OH$, $-COOH$, $-SO_3H$.

The problem of specific adsorption onto the adsorbent surface is virtually eliminated if a non-specific surface can be used. Such a surface is produced by substitution of the surface silanol groups by non-polar organic groups such as octadecyl or other alkyl groups. These so-called reverse phase packings are generally employed with aqueous or alcoholic eluents. They extract the less polar component of any eluent (e.g. an alcohol, acetonitrile etc. from an aqueous mixture) so that partitioning now occurs between an aqueous eluent and a less polar stationary phase.

Surprisingly silica gels modified with non-polar hydrocarbon groups are particularly useful for the separation of large highly polar organic molecules. Generally peak sharpness is enhanced if the compound can be ionized and paired with some counter ion to give an ion-pair which will be extracted into the non-polar stationary phase.

An extremely successful form of ion pair chromatography which we have recently discovered employs a detergent as counter ion for the ion pair (16). If, for example, we wish to separate amine cations, 0.1% w/w of sodium lauryl sulphate may be added to the eluent. With a hydrocarbon bonded packing, the large lauryl sulphate anions are strongly adsorbed at the hydrocarbon surface to give what amounts to an *in situ* ion exchanger which is ideal for retention of amine cations. Figure 2 shows such a separation of tricyclic antidepressants (17), and illustrates the excellent resolution and peak sharpness which can be obtained.

Other new materials have also been developed (18), and in particular new ion exchangers where the ion exchanging

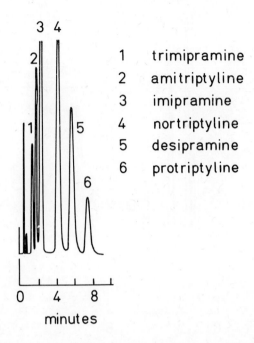

Fig. 2. Separation of tricyclic tranquillizers by reverse phase ion-pair partition chromatography. Column packing 6 µm SAS-silica (WLCU); column 125 x 5 mm; eluent methanol-water-ammonia-sodium lauryl sulphate 75:25:0.5:0.1 v/v; detection UV photometer at 240 nm. Solute identification: (1) trimipramine, (2) amitriptyline, (3) imipramine, (4) nortriptyline, (5) desipramine, (6) protriptyline. (Reproduced by permission of J. Jurand and J. Chromatog. from Ref. 17).

group is bonded to a silica surface. Typical groups are $-Si-(CH_2)_3NH_2$ and $Si-(CH_2)_2-\langle\!\!\!\bigcirc\!\!\!\rangle-SO_3H$. Since the basic silica structure is rigid these materials can readily withstand high pressures, and since the ion exchange layer is unimolecular, mass transfer is rapid as in a normal silica gel. Rapid ion exchange separations are thus possible, such as the high speed separation of nucleotides shown in Figure 3.

Further new materials which give rapid analysis and high resolution will undoubtedly be developed and we may soon expect that virtually all compounds of interest to chemists, biochemists and clinical chemists will yield to high

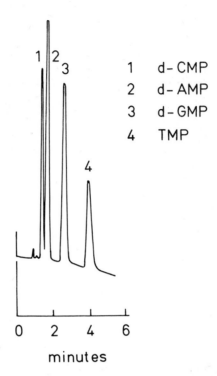

Fig. 3. Separation of DNA nucleoside monophosphates by ion-exchange chromatography. Column packing 6 µm NH_2-Silica (WLCU); column 125 x 5 mm; eluent 0.01 M Na_2HPO_4 + H_3PO_4 (pH 2.46); detection UV photometer at 254 nm. Solute identification: (1) 2'-deoxycytidine-5'-MP, (2) 2'-deoxyadenosine-5'-MP, (3) 2'-deoxyguanidine-5'-MP, (4) thymidine-5'-MP.
(Reproduced by permission of J. Chromatog. from Ref. 17).

performance liquid chromatography.

REFERENCES

1. Huber, J.F.K. and Hulsman, J.A.R.J. (1967). *Anal. Chim. Acta*, 38, 305.
2. Kirkland, J.J. (1969). *J. Chromatog. Sci.* 7, 7.
3. McNair, H.M. and Chandler, C.D. (1974). *J. Chromatog. Sci.* 12, 425.
4. Baker, D.R., Williams, R.C. and Steichen, J.C. (1974).

J. *Chromatog. Sci.* 12, 499.
5. Fleet, B. and Little, C.J. (1974). *J. Chromatog. Sci.* 12, 747.
6. Martin, M., Blu, G., Eon, C. and Guichon, G. (1975). *J. Chromatog.* 112, 399.
7. Knox, J.H. and Vasvari, G. (1973). *J. Chromatog.* 83, 181.
8. Martin, A.J.P. and Synge, R.L.M. (1941). *Biochem. J.* 35, 1358.
9. Kirkland, J.J. (1972). *J. Chromatog. Sci.* 10, 593; (1973) *J. Chromatog.* 83, 149.
10. Halasz, I., Endele, R. and Asshauer, J. (1975). *J. Chromatog.* 112, 37.
11. Persson, B-A. and Karger, B.L. (1974). *J. Chromatog. Sci.* 12, 521.
12. Laird, G.R., Jurand, J. and Knox, J.H. (1974). *Proc. Soc. Analyt. Chem.* 310.
13. Majors, R.E. (1973). *J. Chromatog. Sci.* 11, 88.
14. Knox, J.H. and Parcher, J.F. (1969). *Analyt. Chem.* 41, 1599.
15. Knox, J.H. and Jurand, J. (1975). *J. Chromatog.* 103, 311.
16. Knox, J.H., Laird, G.R. and Jurand, J. (1976). To be published.
17. Knox, J.H. and Pryde, A. (1975). *J. Chromatog.* 112, 171.
18. Pryde, A. (1974). *J. Chromatog. Sci.* 12, 486.

SOME EARLY EXPERIENCE WITH A MOVING WIRE LIQUID CHROMATOGRAPH

H. G. J. Worth

*Department of Clinical Chemistry,
Queen Elizabeth Medical Centre, Birmingham.*

The increased use over the last fifteen years of solid stationary phase columns with a liquid eluent phase for the separation of complex mixtures of chemicals has led to a need for sensitive detectors to monitor the column eluent. It is the development of such detectors over the last decade which has in turn necessitated closer control of column effluent rate by controlling the flow rate by the pressure applied to the column which has resulted in the present technology of high pressure liquid chromatography. One of the first detectors used in liquid chromatography was the Pye System II Liquid Chromatograph. The instrument makes use of a moving wire in its detector system, a principle which is still used in modern intrumentation, and I hope some other contributors to this Symposium will be showing how many of the earlier problems of this technique have now been overcome.

The System II Liquid Chromatograph is shown diagramatically in Figure 1. A fine steel wire passes through the instrument from the feed spool to the collecting spool. Between the spools the wire passes through three ovens and the coating block. The temperature of each oven is set independently of the other two. The oven elements are surrounded by asbestos blocks and where the wire passes through the blocks it is protected by alumino-silicate glass tubing through which it passes. Constrictions at the ends of these tubes enable a slight positive pressure of inert gas to be maintained within the tubes, thus protecting the wire from oxidation at high temperatures. When the instrument is in operation, the wire is slowly wound from the feed spool on to the collecting spool. In so doing it first passes through the cleaner oven where any grease or other contaminants are

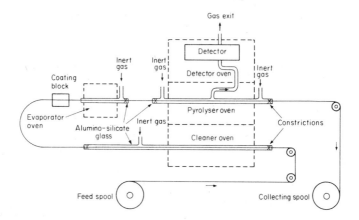

Fig. 1. Schematic diagram of liquid chromatograph.

oxidised off the wire. It then enters the coating block where it passes through the column eluent and becomes coated with eluent. The remainder of the eluent passes out of the coating block and is collected. In the evaporator oven the column solvent is removed. In the pyrolyser oven the column solute is oxidised off the wire, and because the alumino-silicate tubing has an inert gas inlet at both ends, these oxidation products are forced up the central exit into the detector, which is a hydrogen/air flame ionization detector. The presence of any pyrolysis products are thus detected, by connecting a pen recorder to the F.I.D. output.

The efficiency and sensitivity of the instrument as a detector are dependent on a number of variable factors, e.g. the rate of elution of the column, the speed of travel of the wire (which is variable) and the temperatures of the various ovens. As the column eluent rate is decreased, the detector becomes apparently more sensitive as more wire passes through the coating block for the complete elution of any one fraction. However, the fractionation procedure becomes that much slower and appreciable amounts of solute may be pyrolysed in the detector making the yield from the column less quantitative. By increasing the wire speed the detector becomes apparently more sensitive as more wire is coated by a given solute fraction. It is important that the oven temperature settings are determined according to the analysis in hand. The evaporator oven should be set high enough to remove all solvent and, ideally, no solute. The pyrolyser oven should be high enough to remove all solute so that the wire is not contaminated when it is rerun through the system. It is recommended that the cleaner oven be set at least 150°C higher than the pyrolyser oven so that no interference from contaminants is detected.

In this present study the instrument was used for the detection of lipid class fractions separated on silicic acid columns. Columns of silicic acid (1 g) in a petroleum ether (40-60°C) suspension were packed in 5 mm diam. glass columns, and eluted successively with increasing amounts of ether in petroleum ether. Cholesteryl esters were eluted at 1% ether, triglycerides at 4%, non-esterified fatty acids at 8%, free cholesterol at 10%, diglycerides at 25% and monoglycerides at 100% ether. 100% methanol was used to elute phospholipids. The following settings for the instrument were used in these analyses:

Inert gas flows (oxygen free nitrogen)
    pyrolyser oven   30 ml/min
    cleaner oven   20 ml/min
    evaporator oven   5 ml/min
F.I.D.
    hydrogen flow rate 30 ml/min
    air flow rate   500 ml/min
Oven temperature
    pyrolyser   350°C
    cleaner   600°C
    evaporator   50°C
Wire speed minimum (ca. 2 in/sec)

These conditions were found to be suitable for all the types of lipid analyses that were carried out. A synthetic lipid mixture containing cholesteryl palmitate, tripalmitin, palmitic acid, cholesterol, diolein, monopalmitin and lecithin dissolved in petroleum ether was eluted from the column as previously described. The elution pattern was followed on the chart recorder and as soon as one fraction was eluted the solvent polarity was increased and the next fraction eluted. Recovery of each constituent was determined by colorimetric analysis using the Bloor method for cholesterol, Duncombe's method (2) for fatty acids, and glycerides after hydrolysis and the Fiske and Subbarow method (3) for phospholipids after digestion. Recoveries ranged from 93.4% to 99.5%. The response time from when the wire first becomes

TABLE I

Serum Lipid Analysis on a Number of Patients
(Results are expressed in mmol/l)

Fraction	Reference Range (Mean ± S.D.)	D.M.	M.P.	T.V.	J.W.	M.I.	A.D.
Cholesteryl Esters	2.76 ± 0.28	5.85	5.24	3.44	5.42	2.88	0.96
Triglycerids	1.40 ± 0.28	2.42	2.61	1.06	10.47	2.08	3.50
N.E.F.A.	0.302 ± 0.142	0.095	0.299	0.900	1.44	0.577	0.419
Free Cholesterol	1.04 ± 0.10	2.20	1.84	1.08	2.30	0.926	1.36
Diglycerides	0.208 ± 0.130	0.275	0.160	0.165	0.387	0.161	0.171
Monoglycerides	0.300 ± 0.148	0.192	0.363	0.247	0.520	0.293	0.267
Phospholipids	3.83 ± 0.84	4.46	4.68	4.12	7.20	2.76	3.97

D.M. Nephrotic syndrome
M.P. Nephrotic syndrome
T.V. Diabetes with ketosis/acidosis
J.W. Hyperosmolar non-ketotic diabetic (at time of analysis (blood sugar) >50 mmol/1)
M.I. Lactic acidosis (at time of analysis (HCO3−) 4.0 mmol/1 (blood sugar) 12.2 mmol/1, (Lactic acid) 12.2 mmol/1
A.D. Diabetic (at time of analysis (blood sugar) 46.2 mmol/1)

coated with solute to the initial response of the pen recorder is approximately 5 seconds.

This analysis procedure has been applied to the investigation of lipids extracted from human serum and in animal experiments from rat serum and tissues such as adipose, liver and kidney. In some cases incomplete resolution of nonesterified fatty acids from cholesterol and triglycerides occurred. This was apparent from the elution pattern and was overcome by extracting the nonesterified fatty acids into an alkaline aqueous fraction, acidifying and re-extracting into an organic solvent.

Serum lipid analyses are typified by the results shown in Table I. This undoubtedly is the area of greatest contribution to clinical chemistry, but it must be borne in mind that this is a time-consuming procedure, and is by no means a routine analytical tool in its present form. The main problem lies in the fact that the analysis time is long compared to the separation time. If the peak heights or areas of the elution pattern obtained on the pen recorder could be shown to be quantitative, then additional analysis time would not be needed and the procedure would be of considerable use to the clinical chemist. Even with careful measurement of column elution rates and peak areas on the elutions patterns, a method of quantitation could not be found. A number of factors must be contributory to this; namely insufficient control of column flow rate; the method of pyrolysis was poorly controlled so that the F.I.D. response may be variable; and finally the wire speed was also poorly controlled. Of these points the first is undoubtedly the most important, and it is in this area where we have seen the greatest improvement with the production of finely controlled high pressure columns.

In conclusion, I hope this contribution has shown in principle how a moving wire detector has potential in high pressure liquid chromatography, and that some of the other contributors to this Symposium will be able to show how some of the initial problems have been overcome.

Acknowledgement
I wish to thank Pye-Unicam Limited for permission to reproduce Figure 1 from the System II Liquid Chromatograph technical manual.

REFERENCES

1. Bloor, W.R. (1916). *J. Biol. Chem.* 24, 227.
2. Duncombe, W.G. (1963). *Biochem. J.* 88, 7.
3. Fiske, C.H. and Subbarow, Y. (1925). *J. Biol. Chem.* 66, 373.

TRACE ENRICHMENT IN LIQUID CHROMATOGRAPHIC ANALYSIS

P. Williamson

*Waters Associates, Vauxhall Works,
Stockport, Cheshire.*

INTRODUCTION

Liquid chromatography has been developed primarily as an analytical technique and emphasis has been placed on separations where sufficient samples was available to permit quantitation with good precision. However, accurate trace analysis necessitates the operation of liquid chromatographs at very low detection limits and this has recently become of prime concern in environmental chemistry, pollution analysis and in drug metabolism research. In the past, it has been concluded that liquid chromatography was insensitive because of the detection system being used, but there are at least four major interrelated areas which can adversely affect the ability to measure accurately trace quantities:
1. Sample preparation
2. Column performance
3. Injection volume
4. Detector(s) employed.

The over-all detection limit of the system is not the minimum concentration in the detector but the minimum amount (mass) of a component which can be injected and still detected, and this is determined by the product of the concentration of the component in the sample and the effective injectable volume of sample. Restriction of either of these contributing factors will limit the system sensitivity. Where sample concentration is low but the available sample volume is not limited, trace enrichment may be utilised to achieve lower detection limits.

Method

Bonded phase packing materials for liquid chromatography

generally have medium to low surface polarity, the bonded functional groups being normal hydrocarbons, phenyl groups, propylamine groups etc. The mobile phases used with these packings are usually quite polar, e.g. methanol:water, acetonitrile:water. These non-polar or medium polarity packings have a strong interaction with molecules that are non-polar or moderately polar. The polarity balance between the mobile phase and the stationary phase is regulated during development of the separation. As the organic solvent portion of the mobile phase increases it effectively competes with the stationary phase for the sample components and their retention volumes on the column are lowered. Conversely with no organic solvent in the mobile phase the sample molecules will have essentially zero mobility because of their strong preference for the stationary phase. In fact the instant the molecules contact the stationary phase they are completely adsorbed and will remain on the stationary phase until the organic portion of the mobile phase is raised high enough to begin competing with the stationary phase for the adsorbed sample components. Due to the high surface area of these stationary phases, they have the capacity to hold significant quantitaties of adsorbed materials.

Adsorbed materials may be selectively removed by continuously changing the mobile phase composition with solvent programming. The molecules can be selectively adsorbed initially depending on the packing material used, the nature of the molecules to be analysed and the condition of the mobile phase. The combination of these two concepts and the proper manipulation of the variables available form the basis of the trace enrichment technique. Clearly if the molecules are adsorbed in a narrow band at the top of the column and are selectively removed later then this is an extremely powerful method of concentration.

The criteria which must be met in order to successfully apply trace enrichment as a sample injection technique are that the sample components of interest must:
1. Display a "k'" of essentially infinity in solvent A.
2. Display a "k'" of almost zero in solvent B.
   (Note that solvents A and B must be miscible)
3. Be adequately resolved at a given solvent composition of A and B.

The advantages of trace enrichment technique over classical concentration methods (e.g. solvent extraction, carbon adsorption) are:

1. Concentration and analysis can be done on the same column, thus eliminating errors due to handling losses.
2. Extraction of the sample components from water can be made 100%.
3. Fast sample handling.
4. Simple method for providing enough sample for structure analysis.
5. Columns are conveniently and rapidly cleaned for multiple analysis.
6. Reproducibility and control make it suitable for standard methods.

It is of prime importance that the solvents employed in trace analysis are of high purity. In particular the water used in trace enrichment technique should first be passed through the same column that is used for the trace enrichment.

Conclusion

The trace enrichment technique described can be readily applied to trace analysis where sample concentration is low but not limited in volume. It can also be used to prepare in a pure state sufficient of any of the sample components for identification and structure analysis. This can often be accomplished with a single trace enrichment separation; if necessary multiple runs can be made and the fractions combined, as long as the solvent programming is reproducible.

THE USE OF SOFT PACKING MATERIALS OF SMALL PARTICLE SIZE IN THE SEPARATION OF BASIC PROTEINS BY HPLC

Sj. van der Wal and J. F. K. Huber

*Laboratory for Analytical Chemistry,
University of Amsterdam.*

ABSTRACT

Conditions for HPLC separations of basic proteins with a molecular weight between 5,000 and 50,000 were investigated to provide a quantitative analytical as well as preparative tool for biochemical and medical research.

Oxidised cytochrome c (pI 10.4) was chosen to serve as a model compound for these proteins being stable and easily detectable.

Ion-exchangers with a polystyrene-divinylbenzene or cellulose matrix were found to be not useful.

Better results were obtained with hydroxylapatite and (met) acrylic based cation exchangers.

The efficiency of these packing materials and its dependence on particle-size in separating cytochrome c-like compounds is demonstrated. The chromatographic behaviour of oxidised cytochrome c was investigated varying the cation concentration, the pH of the eluent and the temperature.

Some applications in the biochemical and medical field are shown.

This paper has been submitted for publication in full to *Anal. Biochemistry.*

RAPID ION-EXCHANGE CHROMATOGRAPHY OF PLASMA AMINOACIDS

R. S. Ersser

*Department of Chemical Pathology,
The Hospital for Sick Children and
Institute of Child Health,
London.*

Biological fluids such as plasma contain a variety of aminoacids in addition to those used in protein synthesis. Satisfactory separation of these mixtures requires more complex analytical conditions than those suitable for the investigation of protein hydrolysates. These result in correspondingly slower analysis times, which has restricted the use of earlier ion-exchange methods for plasma aminoacids to the analysis of samples from a few highly selected patients. Before these techniques can be considered suitable for "routine" clinical purposes, accurate, reliable, sensitive and convenient methods must be devised which are economic in both labour and materials.

A fully automated instrument is necessary to achieve these ideals (1) but in general the resolution, speed and sensitivity of the analysis is primarily influenced by separation parameters such as the chromatographic properties of the ion-exchange material, the composition and flow rate of the elution buffers, and the dimensions and temperature of the column.

The aim of these studies has been to determine the significance of these factors and their contribution to the quality of the analysis. A practical compromise between the superb but slow high resolution techniques and the rapid methods for simple mixtures such as protein hydrolysates was sought. A method was required which would be generally useful for paediatric purposes such as the investigation of patients with aminoacid disorders and the monitoring of their dietary treatment.

RESULTS AND DISCUSSION

Ion-Exchange Resins

Uniform spherical beads of cation-exchange resin, less than 15 µm diameter, for which 7-8% (by weight) divinyl benzene (dvb) was used for polymerization of the styrene monomer (cross-linking), are the most suitable for rapid aminoacid analysis (2). Lower cross-linked resins (e.g. 4% dvb) distort under the pressures which result from using fast flow rates with small diameter beads. Higher cross-linked resins (e.g. 12% dvb) have greater mechanical strength but require comparatively longer times for complete exchange of ions in the denser bead matrix and lack the discriminating power of the 8% dvb beads. Incomplete or irregular sulphonation (3) and the relative proportions of the isomers of dvb used for polymerization (4) affect the performance of the resin.

The significance of these variations in resin characteristics increases with the complexity and composition of the mixture being resolved. Protein aminoacids can be eluted from the resin column in three major groups of resolved compounds using three buffers of differing pH and salt concentration in sequence. Appropriate changes in pH, salt concentration, and flow rate of the buffers, or a minor alteration in the column temperature will usually compensate for differences in the resolving properties of nominally similar resins. This type of analytical manipulation is either less successful or results in protracted analysis times when applied to the separation of physiological fluids. The large number of pairs or groups of aminoacids which partially overlap in these mixtures often means that in order to achieve adequate resolution in one area of the chromatogram, that of another is sacrificed.

Resins between 7 and 7.5% cross-linked are the most satisfactory for the separation of the acidic and neutral aminoacids present in physiological fluids especially when using lithium buffers (2,5). The separation of the asparagine, glutamic acid, glutamine area of the chromatogram (Fig. 1a) is superior to that achieved with 8% resins (Fig. 2). Citrulline can be separated from alanine without the need to reduce the pH of the buffer after the elution of glutamine, which increases analysis times and broadens the peaks (5). In contrast, when using stepwise changes in elution buffer, greater difficulty is experienced in separating cystathionine from

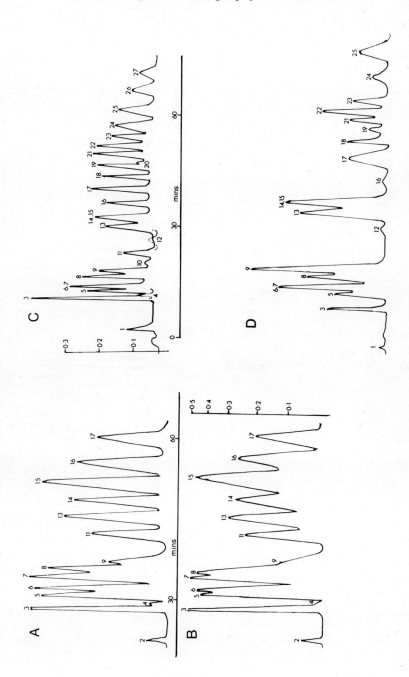

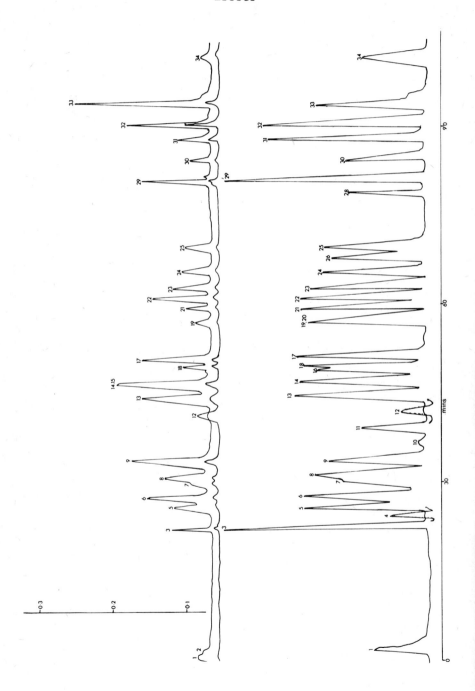

Fig. 1. Separation of Acidic and Neutral Aminoacids.

A and B Effect of Colorimetry on Resolution
Standard mixture of aminoacids separated on identical 300 x 8 mm column of Ostian 0802 resin using 0.3 M lithium buffer, pH 2.75, containing 6% 2-methoxy-ethanol, at a flow rate of 70 ml/hr and column temperature of 37°C.

A: Nitrogen-Segmented Colorimetry
B: Capillary Reactor

C and D Comparison of Gradual and Stepwise Changes in Buffer Aminoacids separated on identical 220 x 4 mm column of C3 resin at a flow rate of 25 ml/hr, and column temperature of 42°C. Initial buffer for both: 0.3 M lithium, pH 2.70, 6% 2-methoxy-ethanol.

C: Standard mixture of Aminoacids. After 24 minutes pH 4.50 buffer mixed with pH 2.70 in chamber (volume 12 ml) to produce gradual rise in pH (1).
D: Normal Plasma (65 µl). After emergence of valine (no. 17), second buffer pH 4.15 used stepwise.
(See Key for identity of peaks)

Fig. 2. Gradient Elution of all Aminoacids on a Single Column.

Aminoacids separated on a 350 x 2.6 mm column of Rank-Hilger resin at a flow rate of 10 ml/hr. Gradient generated by mixing pH 2.20 with pH 11.50 buffers (chromaspek instrument). Temperature elevated from 41-60°C and 2-methoxy-ethanol buffer replaced after emergence of alanine.
Upper trace 40 µl normal plasma; lower trace standard mixture.
(See Key for identity of peaks)

methionine. Organic solvents added to the early buffers to improve the resolution of serine from threonine and of the proline, glycine, alanine, citrulline group, may soften the lower cross-linked beads leading to higher operating pressures after only 2 or 3 analytical cycles. This bead distortion could be greatly reduced if a 2 M lithium citrate buffer pH 4 was pumped onto the column whilst packing the resin, and carefully locating the top column fitting on the resin surface to restrain subsequent swelling of the beads.

The reduced solid-phase mass transport resistance and

KEY

Identification of Compounds Numbered in Figures 1 and 2

No.	Compound	No.	Compound
1	Taurine	18	Cystine
2	Urea	19	Methionine
3	Aspartic acid	20	Cystathionine
4	Hydroxy-proline	21	Iso-Leucine
5	Threonine	22	Leucine
6	Serine	23	nor-Leucine
7	Asparagine	24	Tyrosine
8	Glutamic acid	25	Phenylalanine
9	Glutamine	26	Homocystine
10	Sarcosine	27	β-Amino-iso-butyric acid
11	α-Amino-adipic acid	28	γ-Amino-n-butyric acid
12	Proline	29	Histidine
13	Glycine	30	Tryptophan
14	Alanine	31	Ornithine
15	Citrulline	32	Lysine
16	α-Amino-n-butyric acid	33	Ammonia
17	Valine	34	Arginine

lower diffusion effects of micro-bead resins (less than 10 μm diameter) allow faster exchange, as the chromatographic system operates closer to equilibrium conditions. These improvements can be used to increase the resolution of long columns (6) or they can permit adequate separation on shorter columns at faster flow rates, increasing both speed of analysis and sensitivity (7). The separation of early eluted amino acids shown in Figure 1a took twice as long if performed on a 50 cm long column of Ostion LGKS 0803 (14 μ). By using even smaller beads (Rank Hilger, 8 μ), the analysis time could be further reduced (Fig. 2) to about a quarter of that necessary for the popular 14-16 μm diameter resin particles (2).

Buffer Composition and Elution Profile

Aminoacids were first successfully separated on cation-exchange resin columns using acidic sodium citrate buffers of increasing pH and sodium concentration either sequentially in a stepwise manner or as a continuously changing gradient. Some investigators have preferred to elute all aminoacids from a single long column, whilst others have used a long column for the separation of acidic and neutral aminoacids and a second shorter column eluted with higher molarity buffer to separate the firmer held, more basic compounds (3). Lithium citrate buffers (2) were shown to be superior for the separation of acidic and neutral aminoacids (especially asparagine and glutamine), but at higher molarity were less satisfactory than sodium containing ones for the rapid separation of basic aminoacids (1).

Gradient elution was shown to be superior to stepwise changes in buffer composition for the chromatography of acidic and neutral aminoacids (1) as illustrated in Figures 1c and 1d. The pH, rather than salt, gradient system of Thomas (8) for complete analysis on a single column was initially unsuitable for physiological fluids as a satisfactory basic buffer composition to replace insoluble lithium phosphate had not been devised. The recent introduction of alkaline buffers containing borate ions (9,10) has overcome this limitation and has allowed the type of separation illustrated in Figure 2 to be obtained.

The "escarpment" shaped interruptions to the recorder baseline exhibited when stepwise changes are used (Fig. 1d) are eliminated with gradient elution (1). Odd shaped frontally eluted peaks which are difficult to integrate (e.g. cystine, Figure 1d) are also avoided. The rise of the baseline in the latter part of the chromatogram, associated with the use of high molarity buffers does not occur when solely pH changes are employed (8,10) and column regeneration is more rapid.

Colorimetry

Reaction with ninhydrin is still the most useful detection system for mixtures which contain aminoacids in addition to those present in proteins. Trinitrobenzene suplhonic acid does not react with proline, only reacts poorly with non α-aminoacids, and lacks the long term stability necessary for automation (Barlow and Ersser, Unpublished). Fluorimetric reagents, such as o-phthalaldehyde and flurescamine, offer increased sensitivity but do not react with all possible aminoacid structures without preliminary complex chemistry on the column effluent.

The primary disadvantages of ninhydrin are: the need for a reducing agent, the need to keep the system free of oxygen during reaction, and the slow reaction rate. The resolution of aminoacids must be maintained during reaction which may take 8-15 min even at 100°C. This was originally achieved by performing the reaction in a long teflon capillary coil (11). Hrdina (7) showed that this resulted in appreciable spreading of the recorder peaks. Two methods have been devised to minimise this effect. The first is to reduce reaction time (to 90 seconds) by elevating the incubation temperature to 150°C. This system is used in the D 500 instrument (Durrum Instrument Co., California, U.S.A.) but requires a sophisticated pressurised reaction coil to prevent the reagents boiling and uncontrollable liquid expansion.

A simpler solution is to segment the reagent stream with gas bubbles (7). The magnitude of this effect is shown in Figures 1a and 1b, and the amount of peak spreading caused by the capillary coil reactor is constant (7), precluding this type of system from use with narrow bore columns eluted at slow flow rates. Electronic decoding of gas and liquid segments eliminates the mixing which occurs with mechanical de-bubbling and improves detector response (12).

Sample Loading

Separation of the acidic and neutral aminoacids least well retained by the resin, was influenced by the method of automatic samples loading employed (13) and the pH of both sample and loading buffer needed optimising for each particular instrument. If the pH of the loading buffer was too low, the peaks in the early part of the chromatogram (Nos. 1-9, Figs. 1 and 2) were crowded together and poorly resolved. If the pH was too high, excessively broad and equally poorly separated peaks resulted. Cysteic acid, phosphoethanolamine and taurine were only mildly retarded during their passage down the resin column. They experienced considerable spreading when loaded through the high pressure pump; could be washed off the Technicon TSM resin cartridge during application if sample volumes greater than 20 µl were loaded; and were therefore generally ignored.

Complete Procedures

Two rapid (2 hr) procedures for the analysis of plasma

aminoacids were devised as a result of these studies. Both use 8% cross-linked resin as until recently the author had been unable to obtain 7.5% beads of 8 μm diameter.

Using the TSM, acidic and neutral aminoacids were separated on a 220 x 4 mm column of C3 resin as illustrated in Figure 1c. A simplified version of the gradient system of Ersser and Seakins (1974) was used. Basic aminoacids were separated on a second 220 x 4 mm column of C3 at 57°C using a 0.5 M sodium citrate buffer pH 4.3, followed by a 1.0 M sodium citrate buffer pH 6 to elute arginine (1). A superior, single column, gradient elution technique was developed in co-operation with C. Murren and D. Stelling of Rank Hilger on their Chromaspek instrument and is illustrated in Figure 2. Neither method adequately separated asparagine or citrulline but most other compounds could be resolved sufficiently for meaningful identification and quantitation.

Precision and Quantitation

The emergence times of aminoacid peaks separated by rapid methods had co-efficients of variation (CV) of less than 1%, which are similar to those obtained from longer analytical cycles (14). There was no significant difference in reproducibility between results obtained from either stepwise or gradient elution.

The electronically integrated areas (1) or manually measured heights (14) of peaks resolved almost to baseline were precise within a CV of less than 2%. The precision of partially overlapping doublets depended on the relative proportions of each compound; the poorest was for those aminoacids first eluted from the TSM column (Fig. 1c) which yield CV's between 3.5 and 5%. Recovery of added standards was similarly variable between 95-103%, depending on the degree of resolution and stability of compounds loaded into the samplers awaiting analysis (14).

Conclusions

If used in conjunction with a preliminary thin-layer chromatographic assessment of aminoacid content, and provided doubtful results are confirmed by high resolution techniques, rapid ion-exchange chromatography of plasma aminoacids can be an extremely useful and time-saving technique of quantitative estimation. It is suitable for most clinical purposes and in the author's laboratory, these methods perform 80% of the quantitative aminoacid analysis required for the diagnosis and treatment of children with aminoacid disorders. Modern aminoacid analysers have the necessary reliability for continuous routine operation.

REFERENCES

1. Ersser, R.S. and Seakins, J.W.T. (1974). *In* "Automation in Analytical Chemistry, 1971", pp. 12-21. Technicon Instrument Co. Ltd., Basingstoke, England.
2. Benson, J.V. (1971). *In* "New Techniques in Aminoacid Peptide and Protein Analysis", pp. 3-42. (A. Niederwiser and G. Pataki, eds.) Ann Arbor Science Publishers, Michigan, U.S.A.
3. Hamilton, P.B. (1966). *In* "Advances in Chromatography", Vol. 2, pp. 3-62. (D. Giddings and A.P. Keller, eds.) Dekker, New York.
4. Long, C.L. and Geiger, J.W. (1969). *Anal. Biochem.* 29, 265-283.
5. Ersser, R.S. (1973). *Lab. Equip. Dig.* 11, 99-105.
6. Scott, C.D. (1972). *In* "Advances in Clinical Chemistry", Vol. 8, pp. 1-41. (D. Giddings and A.P. Keller, eds.) Academic Press, London.
7. Hrdina, J. (1968). *In* "6th Colloquium in Aminoacid Analysis", pp. 60-79. (H. Holy, ed.) Technicon International Division, Geneva.
8. Thomas, A.J. (1970). *Proc. Nutr. Soc.* 29, 108-110.
9. Murren, C., Stelling, D. and Felstead, G. (1975). *J. Chromatog.* 115, 236-239.
10. Ersser, R.S. (1975). *J. Chromatog.* 115, 612-615.
11. Spackman, D.H., Stein, W.H. and Moore, S. (1958). *Anal. Chem.* 30, 1190-1198.
12. Thomas, A.J. (1975). *In* "The Medical Technologist", Vol. 5, No. 2, pp. 32-34.
13. Ersser, R.S. and Mossman, T.G. (1975). *Lab. Pract.* 24, 741-743.
14. Ertingshausen, G. and Adler, H.J. (1970). *Amer. J. Clin. Path.* 53, 680-691.

ANALYSIS OF SERUM LIPIDS AND LIPOPROTEINS BY HIGH SPEED LIQUID CHROMATOGRAPHY

*T. Carter, H. N. Magnani** and Rodney Watts**

*Wolfson Research Laboratories,
Queen Elizabeth Medical College, Birmingham.

**Department of Experimental Pathology,
The Medical School, Birmingham.

INTRODUCTION

The liquid chromatography of lipids has been the subject of a vast area of research and a good review by Aizetmüller (1) has recently appeared. The lack of suitable chromophores in most lipids makes the usual UV detector unsatisfactory and the moving wire-flame ionization detector, described by Dr. Worth earlier in the symposium, is the detector of choice. Worth and MacLeod (2) used this detector to monitor the elution of human plasma or serum neutral and phospholipids from a silicic acid column. Two other methods for the analysis of non-polar serum lipids by high speed liquid chromatography have been published (3,4).

Though the intact lipoproteins can be partially separated by gel permeation on Sepharose 4B, no high speed separations have yet been reported. In this paper we present some results on the analysis of the serum neutral lipids using a single solvent system and flow programming, serum phospholipids using another single solvent system, and serum lipoproteins by the direct application of serum to a fast flow gel column.

Materials and Methods

Columns

Silica Gel H (Merck) was activated at 110°C for 1 h, and glass columns of length 25 x 0.22 cm were dry packed.

A 25 x 0.22 cm glass column, and a 10 x 0.4 cm stainless steel were packed with a slurry of Sepharose CL-4B (Pharmacia) under slight pressure. A 10 x 0.4 cm glass column was packed with a slurry of a mixed polyacrylamide and sepharose gel (ACA22, LKB).

Elution Systems

Neutral lipids were separated on one of the silica gel H columns with 20% diethyl ether in heptane or hexane at a flow rate of 0.5 ml/min until the cholesteryl esters and triglycerides had been eluted, and then at a flow rate of 2 ml/min until free cholesterol had been eluted. Phospholipids (mainly phosphatidyl choline and sphingomyelin) were separated using a second silica gel H column and a solvent system of chloroform-methanol-water 65:25:4 v/v at 0.8 ml/min. Intact serum lipoproteins were partially separated in either single column runs on Sepharose CL 4B or on tandem columns of Sepharose CL 4B followed by ACA22. The solvent system was 0.9% saline.

Lipid Extraction

When only the neutral lipids were of interest 1 vol serum was taken up into 5 vol Dole reagent (isopropanol-heptane-M sulphuric acid 80:20:2 v/v). The mixture was shaken and left 15 min. After a further addition of 2 vol water and 3 vol diethyl ether, the mixture was again shaken, left for a further 15 min and then 5-20 µl (*ca*. 15 µg total lipids) of the *upper* organic phase was injected onto the column.

Phospholipids as well as the neutral lipids were extracted by the addition of 9 vol chloroform-methanol 1:1 v/v to 1 vol serum. The mixture was well shaken and after 15 min 5 vol 3 M sodium chloride were added. The two phases were allowed to separate and 5-20 µl of the *lower* chloroform rich layer were injected onto the second column.

LC Systems

A Pye model LCM 2 liquid chromatograph fitted with an LC 20 pump and a moving wire-flame ixonization detector was used. The 25 x 0.22 cm glass columns were also supplied by Pye-Unicam. The 10 x 0.4 cm stainless steel column was constructed in the departmental workshop. For the lipoprotein separations a Pharmacia peristaltic pump P3 was employed.

Reagents

All organic solvents and reagents (BDH) were of AR grade.

Standard

All neutral lipid standards were > 99% pure. BDH Ltd. supplied cholesteryl stearate and oleate, free cholesterol and palmitic acid. Sigma Chemical Co. Ltd. supplied triolein as well as the phospholipid standards, sphingomyelin and phosphatidyl choline.

Ultracentrifugation

1.5-3.0 ml serum was spun in a density gradient by layering between saline (d = 1.006) and 2 M NaCl after the method of Walton *et al.* (5). A Beckman Spinco ultracentrifuge was employed and the specimens were spun at 40,000 r.p.m. overnight. Very low (VLDL) and low density (LDL) lipoprotein fractions were dialysed against 0.9% saline and then subjected to column chromatography.

RESULTS AND DISCUSSION

Serum Lipids

Figure 1 shows the separation of the neutral lipids found in serum and in Figure 2 the detector response to load is shown. Figure 3 illustrates the separations of serum phospholipids from neutral lipids; also shown is the separation of partially purified serum phospholipids.

The separation of lipid classes by LC has usually been accomplished using sequential or gradient elution techniques and reuse of the columns has involved regeneration steps. However, we have found it far more convenient to use single solvent systems on two separate columns with one increase in flow rate needed for the separation of the neutral lipids. Carotene and vitamin E cochromatograph with the cholesteryl esters and cholesterol respectively. Phosphatidyl choline and sphingomyelin make up approximately 90% of the total serum phospholipids and because of this simplicity of composition no elaborate procedures are required. The small amounts

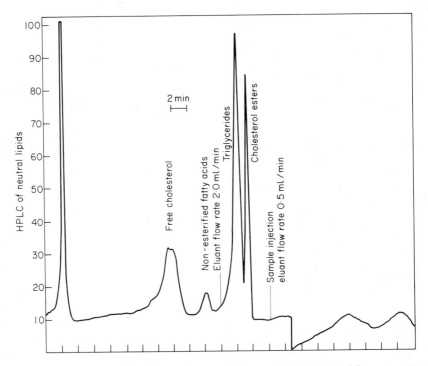

Fig. 1. The separation of ca. 10 µg neutral lipids on a 25 x 0.22 cm column of silica gel H, using 0.5 ml/min 20% diethyl ether in heptane for 6 min, then 2 ml/min. Detector: Pye moving wire-F.I.D., attenuation x 64.

of phosphatidyl ethanolamine and phosphatidyl serine (*ca.* 5%) cochromatograph with the neutral lipids in our system, but lysolecithin can be eluted after the sphingomyelin.

Serum Lipoproteins

Figure 4 shows the separations of whole serum lipoproteins on the 10 cm column of Sepharose CL-4B. The use of the mixed polyacrylamide and sepharose gel, ACA 22, with a claimed exclusion limit of approximately 10^6 Daltons was not so successful, and did not improve resolution of the VLDL from LDL when used as a second tandem column. It also ran more slowly. The exclusion limits for the Sepharose 4 B is about 20×10^6 Daltons, and it would therefore seem feasible that

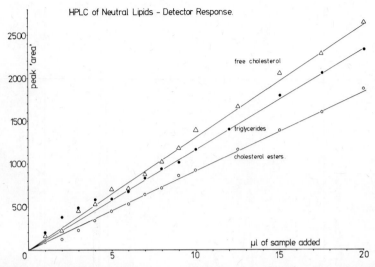

Fig. 2. Detector response to increasing load of some neutral lipids.

further resolution should be obtained on either a single column of Sepharose CL 2B, which has an exclusion limit of about 50×10^6 Daltons, or a tandem system with a 2B column followed by a 4B column.

It may be seen that since the detector responded, lipoproteins might well be quantitated by this technique. However, at this preliminary stage it is not possible to report relative responses of the detector to the different sizes of lipoproteins. It was of interest to note that other proteins did not give a detector response in this system. Even when a concentrated albumin solution was applied directly to the moving wire little significant response was observed.

General Comments

Lipid profiles at present comprise total serum triglyceride and cholesterol levels together with the electrophoretic patterns of the lipoproteins. Gas-liquid chromatography (GLC) (6) can be used to measure free and esterified cholesterol as well as triglycerides, and at the same time gives information on the molecular species of the lipid classes. However, this is more information than is required for many purposes, and LC does give free and esterified cholesterol

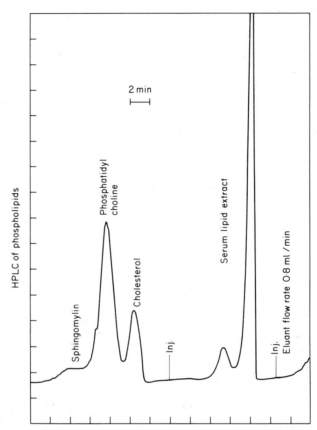

Fig. 3. The separation of ca. 15 µg of total serum lipids or partially purified phospholipids. Column and detector conditions as in Figure 1. The eluent was chloroform:methanol: water 65:25:4 v/v.

results. There are expensive column packings on the market, but for the separation of the dominant serum lipid classes silica gel H is adequate.

The use of silica gel H has also meant that the TLC solvent systems used in our laboratory have been directly applied to the HPLC method.

The most significant aspects of this work are the rapid quantitation of the individual phospholipid classes and the possible direct quantitation of the different serum lipoprotein classes. The phospholipid composition changes from

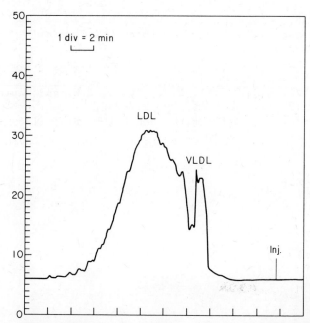

Fig. 4. The partial separation of intact serum lipoproteins on 10 x 0.22 cm Sepharose CL 4B. The eluent was 0.9% saline at 0.1 ml/min and detector attenuation was set at x 32.

health to disease, but relatively little work has been done in this area. There are a number of precipitation and immunochemical methods for the estimation of VLDL and LDL, but at present these are not entirely satisfactory. We do not at this time claim the gel permeation method to be superior, but present indications are hopeful.

Acknowledgements
Pye-Unicam kindly provided their Liquid Chromatography equipment on evaluation loan. LKB and Pharmacia gave us samples of their gels. Mrs. H. Bottom and Miss Stella Taylor gave technical assistance. We also appreciate the interest of Professor K. W. Walton and Dr. P. Wilding.

REFERENCES

1. Aizetmüller, K. (1975). *J. Chromatog.* 113, 231-266.
2. Worth, H.G.J. and MacLeod, M. (1969). *J. Chromatog.* 40,

31-38.
3. Freeman, N.K. (1974). *J. Amer. Oil Chemist's Soc.* 51, 527A.
4. Werthessen, N.T., Beall, J.R. and James, A.T. (1970). *J. Chromatog.* 46, 149-160.
5. Walton, K.W., Scott, P.J., Dykes, P.W. and Davies, J.W.L. (1965). *Clin. Sci.* 29, 217-238.
6. Kuksis, A., Stachnyk, O. and Holub, B.J. (1969). *J. Lipid Res.* 10, 660-667.

SEPARATION OF OLIGOSACCHARIDES

Ten Feizi*, P. Gough* and P. Williamson**

*Division of Communicable Diseases,
MRC Clinical Research Centre, Harrow,
Middlesex.

**Waters Associates, Vauxhall Works,
Stockport, Cheshire.

The usual methods for isolating oligosaccharides from mixtures are time consuming. High pressure liquid chromatography (HPLC) offers promise for the rapid separation of oligosaccharides. The µ Bondapak Carbohydrate column made by Waters Associates has been shown by them to be effective in separating the oligosaccharides of partially degraded starch. We have used two such columns in series to fractionate the oligosaccharides produced by partial acid hydrolysis of dextran. Pharmacia dextran (T2000) was treated with 0.3 N sulphuric acid according to the method of Whelan (1) and 5 mg of the hydrolysate was applied to the high pressure liquid chromatograph (Waters Model ALC/GPC 201) which is equipped with a refractive index detector. In 35 minutes 10 distinct peaks were obtained (Fig. 1). Peak and trough fractions were collected separately and the purity of the peak fractions was assessed by descending paper chromatography. Peaks 1 to 6 were chromatographed for 24 hours in butanol:pyridine:water, 6:4:3 (v/v) and the subsequent peaks at 35:39:26 (v/v) for 40 hours. Considerable purification of the oligosaccharides was achieved (Fig. 2); each peak contained a single major component and only a trace of the oligosaccharide from the previous peak. It should be possible to eliminate the latter by rechromatography. Dextran is largely made up of straight chains of 1→6 linked glucose, with a small proportion of other linkages. Many laboratories are interested in the use of the HPLC system for isolating oligosaccharides for immunochemical and structural studies from more complex polysaccharides such as mammalian glycoproteins. The successful rapid fractionation of such complex structures would make HPLC an invaluable tool.

REFERENCE

1. Whelan, W.J. (1962). *In* "Methods in Carbohydrate Chemistry", Vol. 1, p. 321. Academic Press.

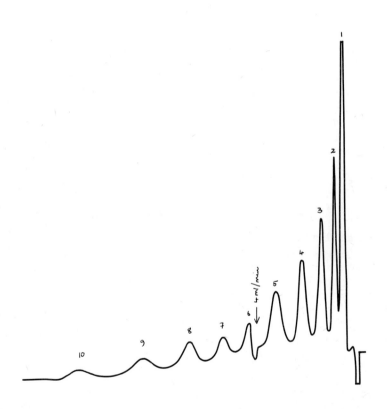

Fig. 1. Chromatography of 5 mg partially hydrolysed dextran on a Waters High Pressure Liquid Chromatograph Model ALC/GPC 201 with a refractive index detector.
Columns: μ *Bondapak Carbohydrate (two in series).*
Solvent: 35% *water,* 65% *acetonitrile.*
Flow Rate: 2 *ml/min* → 4 *ml/min.*

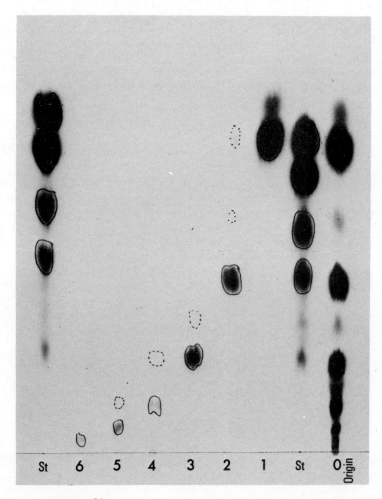

Fig. 2. Descending paper chromatography in butanol:pyridine: water, 6:4:3, running time 24 hours. Symbols: 0 = original dextran hydrolysate; St = standard mixture containing glucose, galactose, maltose and isomaltose (in order of rapidity of migration) and small amounts of higher molecular weight oligosaccharides. The R_G values and probable identities of the main components of the first four peaks were as follows: Peak 1: R_G 1.0 glucose; Peak 2: R_G 0.55 isomaltose; Peak 3: R_G 0.41 isomaltose trisaccharide; Peak 4: R_G 0.3 isomaltose tetrasaccharide. The remaining peaks were not identified.

SEPARATIONS OF SOME STEROIDS, TRITERPENES, GLYCOSIDES, BITTER PRINCIPLES, AND RELATED COMPOUNDS

R. D. Burnett

*Dyson Perrins Laboratory,
University of Oxford.*

These separations have been carried out using a Waters Model M6000 liquid chromatograph equipped with a Waters U6K septumless injection loop.

For some years in the Dyson Perrins Laboratory there has been interest in the micro-biological hydroxylation of steroids and steroid-like compounds. Assay of the products can be a difficult or tedious operation, and for the example shown (Fig. 1), hydroxylation of a tricyclopentadiene derivative, HPLC was the only convenient method of assay. Hydroxy-

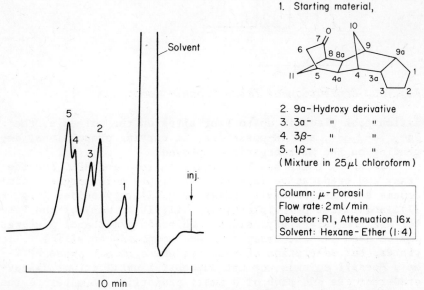

Fig. 1. Assay of Microbiological Hydroxylation Products

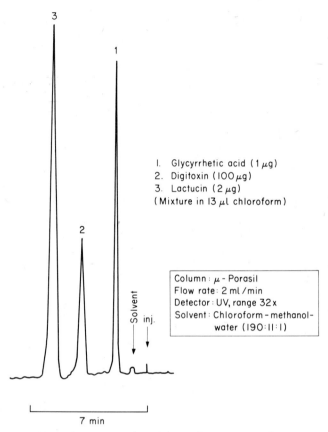

Fig. 2. *Separation of Polar Organic Compounds*

lation took place in up to four sites on the molecule, the 9α-, 3α-, 3β-, and 1β-positions, in varying proportions, depending on the microorganism employed.

Recently there has been interest in using HPLC for the assay and preparative isolation of highly polar organic compounds such as carboxylic acids (e.g. I), glycosides (e.g. II), and bitter principles (e.g. III, IV): many of the latter are themselves glycosides. Chloroform/methanol mixtures can be used for the assay of these compounds on silica t.l.c. plates, but adaptation of this system to normal phase HPLC on μ Porasil columns was not successful until it was discovered that the addition of a small proportion of water to the solvent mixture significantly reduced peak broadening and

Steroids and Glycosides   47

SELECTED REFERENCE COMPOUNDS

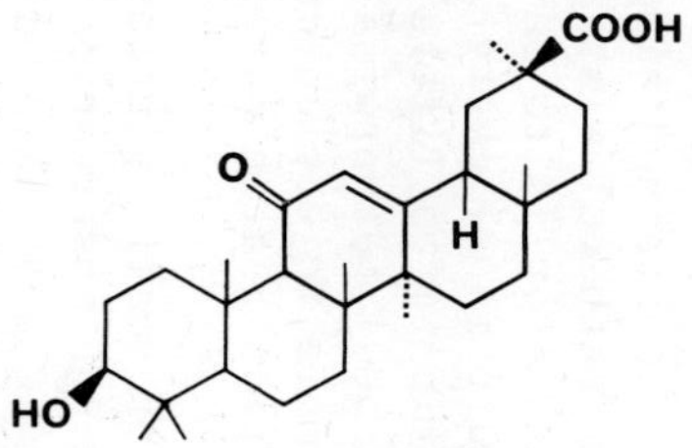

Glycyrrhetic Acid (0·74) (I)

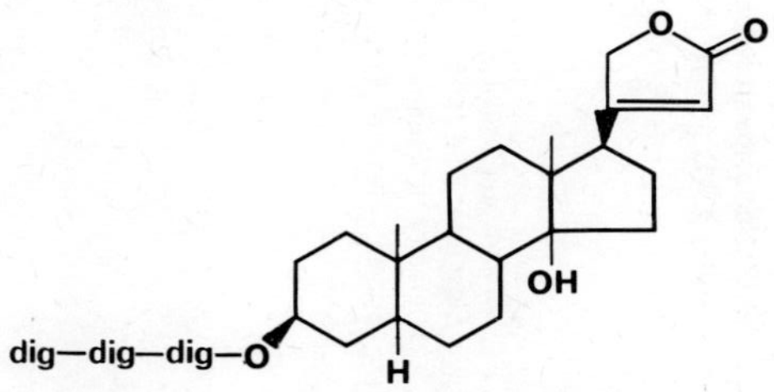

Digitoxin (0·56) (II)

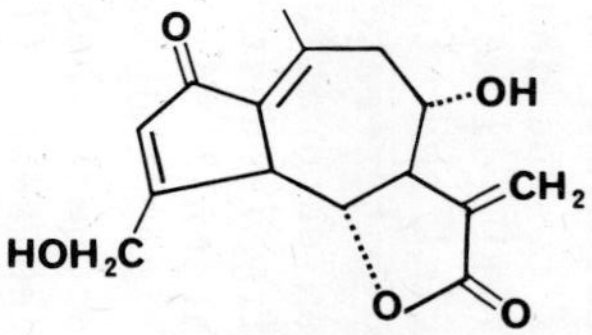

Lactucin (0·47) (III)

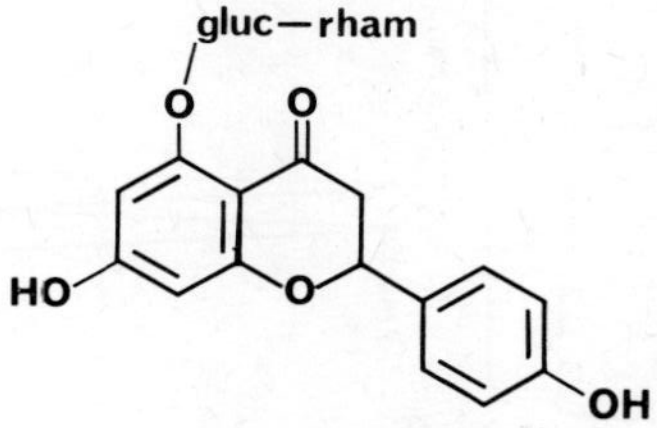

Naringin (0·05) (IV)

Figures in brackets refer to typical R_f values (Silica gel G, Chloroform/Methanol (5:1))

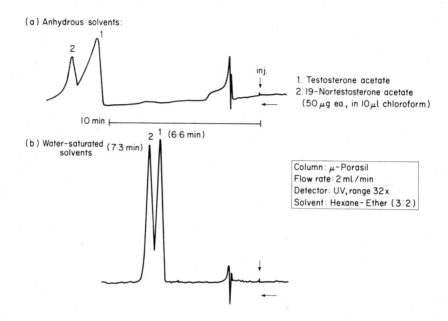

Fig. 3. Separation of Steroids, demonstrating importance of using Water-Saturated Solvents.

tailing (Fig. 2). If a UV detector is to be used a suitable chloroform/methanol/water mixture can be obtained simply by adding water to the desired chloroform/methanol mixture until no more will dissolve. If a RI detector is to be used, it is essential that the solvent mixture is slightly less than totally water-saturated (by addition of a little more methanol when saturation point is reached), otherwise baseline stability is destroyed by the sample injection solvent. This observation leads to the supposition that a partition-type system is set up on the column.

The discovery that the addition of small quantities of water enhanced separations arose from the observation that steroid separations on normal-phase HPLC are enhanced by the use of partially or totally water-saturated solvents (Fig. 3). The degree of water-saturation makes very little difference to separating power or retention times between 50 to 100% water-saturation. However, after using 100% water-saturated solvent for some time it was observed that baseline stability

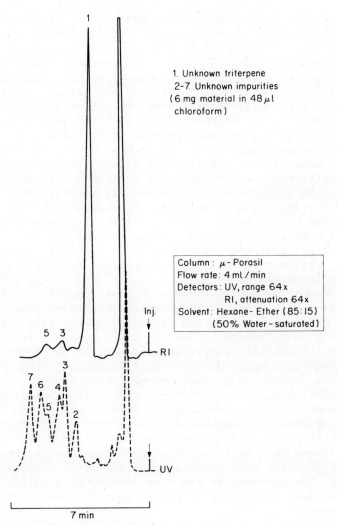

Fig. 4. Preparative Separation.

of a RI detector deteriorates, but is restored by changing to 50% water-saturated solvent. This could perhaps be due to gradual overloading of water on the solid support when using totally water-saturated solvent. As a result, it is recommended that solvents should always be 50-75% water-saturated.

A colleague, Dr. George Hutchinson, has apparently achieved excellent separations of bitter principles using the solvent system 1-butanol:ethanol:water (4:1:5, top layer) on µ Porasil columns. This solvent system is commonly employed for the paper chromatography of sugars: it has not yet been fully investigated for HPLC but initial results suggest that it, and similar solvent systems, warrant further investigation. The solvent system cyclohexane:1-butanol or 2-butanone:water (top layer) may be useful if a less polar solvent is required.

An advantage of normal phase liquid chromatography is that purification and recovery of unknown or valuable samples can be achieved simply and rapidly by fraction collection and evaporation of solvent. Furthermore the system is economical because only a single column of porous silica, such as µ Porasil, with the appropriate solvent system, is required for isolation of compounds of widely differing polarity, and because the solvents employed are readily available and inexpensive.

Preparative separations. µ Porasil, being totally porous has a high capacity for preparative separations. Quantities of several milligrams can be separated in a single injection on µ Porasil columns (4 mm ID x 30 cm) (Fig. 4). For the illustrated example it was required to separate a triterpene (the major peak on the RI detector) from several UV-active impurities present in small proportion. The injection limit, 6 mg, was not determined by peak spreading but by a tendency of the compounds to crystallise in the injection loop and column approach leads. This problem was overcome by changing to a stopped-flow, on-column injector, when it was found that 10-12 mg of mixture could be injected before significant peak broadening of the major component occurred. The tendency of the compounds to crystallise was equally great when using a methanol-water solvent system on reverse-phase packings.

Separations developed on µ Porasil are easily adapted to larger-scale separations using preparative silica columns.

SEPARATION OF OESTROGENS AND DETERMINATION OF OESTRIOL IN HUMAN PREGNANCY URINE

Vera Fantl, C. K. Lim and C. H. Gray

*Department of Chemical Pathology,
King's College Hospital Medical School,
London.*

The oestrogenic steroid hormones are characterised by the presence of a phenolic "A" ring. Oestradiol the human ovarian hormone is the most potent naturally occurring steroid. Oestrone is important in both the biosynthesis and metabolism of oestradiol while oestriol is its major urinary metabolite. Since the isolation of the three classical oestrogens, many other oestrogen metabolites have been found and are still being found in human pregnancy urine. Those included in this present study are oestrone with a 2-methoxyl and 16α-hydroxyl group, oestradiol with a 16 carbonyl group and two other isomeric oestriols, 16 epioestriol and 17 epioestriol.

At present UV absorption is the most sensitive and widely used method for detection of compounds separated by high pressure liquid chromatography. Since the UV absorbing nucleus characteristic of the oestrogens remains intact in their excretion products this eliminates the need to make UV absorbing derivatives for their investigation. Although an absorption maximum occurs at 280 nm, sensitivity of detection of the oestrogens was approximately doubled at 230 nm and trebled at 217 nm. Lower UV wavelengths could not be investigated since significant absorption of the solvent occurred. Throughout this investigation, a Waters Associates Model 6000 solvent delivery with a Model U6K injector, and a Cecil CE 212 detector with variable wavelength adjustment was used.

Using a μ Bondapak NH_2 column (4 mm x 30 cm) consisting of 10 μ diameter silica particles coated with aminopropylsilane, oestrone, oestradiol, oestriol 3-methyl ether, 16α-hydroxyoestrone, 16-oxooestradiol and oestriol were separated

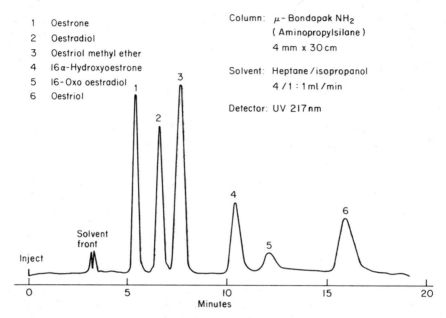

Fig. 1. Separation of Standards.

(Fig. 1). The solvent system was heptane/isopropanol 4:1 v/v with a flow rate of 1 ml/min. Very little separation of 17-epioestriol and 16-epioestriol occurred. These steroids were eluted just before 16α-hydroxyoestrone from which they were not separated. These three steroids have similar structures, the differences occurring in the adjacent oxygen functions present in the D ring.

Using heptane/isopropanol 4:1 v/v oestrone and 2-methoxyoestrone were eluted too soon for any separation to occur. With the less polar combination 19:1 v/v a complete separation was obtained, 2-methoxyoestrone being eluted first. By increasing the flow rate from 1 to 1.5 ml/min the separation was achieved within 10 mins. Using a μ Porasil column (4 mm x 30 cm) consisting of 10 μ diameter silica particles, the following separation was obtained using heptane/isopropanol 37:3 v/v and a flow rate of 1 ml/min: oestrone, oestradiol, 16α-hydroxyoestrone, oestriol 3-methyl ether and oestriol. The order of elution of the steroids is the same as that achieved with μ Bondapak NH₂ except that 16α-hydroxyoestrone is eluted before oestriol methyl ether. However 16-oxooestradiol and 16α-hydroxyoestrone readily separated on μ Bondapak NH$_2$ were only very slightly resolved under these conditions.

Three of the oestrogens when injected on the μ Porasil column under identical conditions as the other standards could not be detected. These were 2-methoxyoestrone, 16-epioestriol and 17-epioestriol. We therefore deactivated the column by equilibration in water saturated heptane in an attempt to reduce the apparent strong absorption of these steroids to the silica particles.

Using water saturated heptane/isopropanol 9:1 v/v, 1 μg 17-epioestriol was chromatographed. A broad tailing peak observed for 17-epioestriol even with a deactivated silica column demonstrated why we failed to detect this compound and 16-epioestriol on the more active silica column. 2-methoxyoestrone with an adjacent phenolic hydroxyl and methoxyl group is even more strongly absorbed to μ Porasil than the epimeric oestriols. 1 μg was not detectable on this deactivated column using the identical conditions. The reason for this unexpected absorption is possibly due to the additional rigidity conferred on the steroid nucleus by the presence of adjacent functional groups in the same geometrical plane. Oestriol with adjacent hydroxyl groups in different geometrical planes does not exhibit this strong absorption.

Reverse phase separation of the oestrogens was obtained using a μ Bondapak C$_{18}$ column (4 mm x 30 cm) with acetonitrile/water 2:3 v/v. Oestriol, 16α-hydroxyoestrone, oestriol 3-methylether oestradiol and oestrone were separated, the reverse order of separation obtained on μ Bondapak NH$_2$. An increase in flow rate of solvent from 1 to 1.5 ml/min after elution of 16α-hydroxyoestrone enabled the total separation to be achieved within 25 mins. Eluted between oestriol and 16α-hydroxyoestrone 16-oxooestradiol was only partially separated from the latter steroid. 16-epioestriol was eluted closely after 16α-hydroxyoestrone as a broad tailing peak while 17-epioestriol with a similar retention time was barely detectable. Even with a faster flow rate of 1.5 ml/min 2-methoxyoestrone was just visible as a hump well separated from oestrone.

Of the three column types we have used, μ Bondapak NH$_2$ produced the best separations. Moreover no losses due to strong absorption were observed. For these reasons this was the column of choice for the study of the oestrogenic hormones.

Huber, Hulsman and Meijers (1) separated oestrone, oestradiol, and oestriol, using normal phase chromatography on diatomaceous earth (particle diameter 28-32 μm) using a ternary system water ethanol 2,2,4-trimethylpentane. Reverse phase chromatography of a large number of oestrogens was carried out by Siggia and Dishman (2) on a terpolymer support (consisting mainly of trifluoroethylene) coated with amberlite LA-1 (n-dodecenal (trialkylmethyl) amine). Using water pH 11.5 (NaOH) with flow programming they separated oestriol, 16-oxooestradiol, 16-epioestriol, oestradiol, 2-methoxyoestrone, and oestrone. The order of elution for 2-methoxyoestrone and oestrone was reversed, compared with our findings using μ Bondapak C_{18} and eluting under neutral conditions with acetonitrile/water. This is probably due to the greater affinity of the more acidic 2-methoxyoestrone for the alkaline eluent than the unsubsituted compound. Using LA-1, a substituted amine as stationary phase these workers observed no strong absorption effects. This is comparable to our results using μ Bondapak NH_2.

During pregnancy, especially in the last trimester, a large increase in urinary oestrogens occurs. Of the total present, oestriol is quantitatively the largest component. Since it is produced mainly by the placenta from precursors supplied by the foetus the determination of urinary oestriol is of value in monitoring the course of pregnancy. Although a small proportion is present as a sulphate conjugate the following method measures only the much larger glucuronide component.

Using 50 μl of a potent β-glucuronidase preparation from E. coli (Boehringer Mannheim) hydrolysis was achieved in 10 minutes at 45°C on a 2 ml sample of urine. A single ether extraction (10 ml), removal of the aqueous layer, drying with Na_2SO_4 and evaporating the ether to dryness, produced an extract suitable for chromatography. The residue was dissolved in isopropanol (200 μl) and 10 μl injected onto the μ Bondapak column. The chromatographic trace produced is shown in Figure 2. A single major peak corresponding to oestriol was obtained. A more polar 3:1 v/v heptane/isopropanol combination enabled oestriol to be eluted within about 10 mins. A similar trace with a large peak corresponding to oestriol was obtained on the μ Porasil column. The minor impurities eluted near the oestriol peak can be eliminated if the ether extract is washed with carbonate buffer before being dried.

Using the same urine sample hydrolysis with ketodase, a β-glucuronidase preparation from beef liver, yielded a simi-

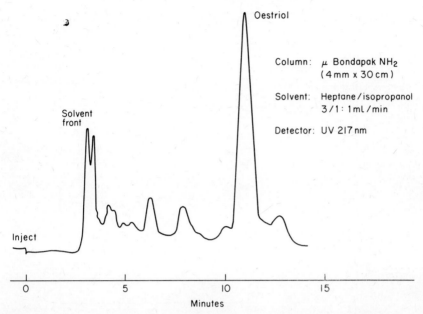

Fig. 2. *Separation of Oestriol from Pregnancy Urine.*

lar concentration of oestriol but hydrolysis had to be performed overnight. Moreover, the speed of the E. coli hydrolysis eliminates the need for a buffer, required for the ketodase hydrolysis. The optimum pH for the bacterial preparation is about 6.5 so that the urine samples need little, if any, adjustment. Acid hydrolysis although faster than ketodase requires at least 30 mins for completion. Under these conditions a lower yield of oestriol was obtained and more impurity was observed at the solvent front.

Routine methods of determination of pregnancy oestrogens rely on the Kober colour reaction with the Ittrich fluorometric end point. More recently developed automated methods eliminate the need for hydrolysis of the oestrogen conjugates and the colour reaction is performed directly on the diluted urine sample. The simplicity of the method makes it very susceptible to interference by other substances, particularly glucose which is often found in the urine of pregnant subjects and has to be eliminated if reliable results are to be obtained.

The present method, although not fully evaluated, is quick. A single measurement can be performed in about 45

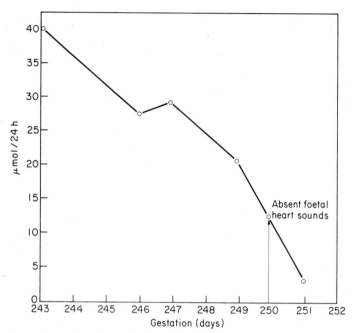

Fig. 3. Association of foetal death with falling urinary oestriol.

mins and it seems likely that interference by other substances will be minimal. With the use of automatic injection for high pressure liquid chromatography the 30 specimens/day possible with manual injections could be significantly increased. Drift problems associated with autoanalyser methods require regular measurement of standards throughout a batch of analyses. The within batch precision of the high pressure liquid chromatogram measured by repeated injection of a urine sample gave a coefficient of variation less than 1%. Linear response was excellent so that only a single concentration of standard needs to be used for quantitation. Without difficulty we have measured about 20 µmole/24 hr of oestriol.

The validity of this method has been demonstrated by measuring serial oestriol excretion in a case of inter-uterine death (Fig. 3). Even before foetal heart sounds ceased, a marked fall in oestriol levels indicated that the pregnancy was in danger.

As early as 1971 Huber *et al.* (1) attempted to measure urinary pregnancy oestrogens by column liquid chromatography.

Using diatomaceous earth as mentioned previously they attempted to measure oestrone and oestradiol as well as oestriol. They used 50 ml urine, acid hydrolysis, three ether extractions followed by carbonate and aqueous washes to obtain an extract for chromatography. Although oestriol was eluted as a single peak after 20 minutes oestrone and oestradiol were eluted with other contaminants. These workers used UV detection in the less sensitive 281 nm region.

Subsequent development of more efficient column packings and the use of low UV wavelengths has now made routine measurement of oestriol excretion during pregnancy, using high pressure liquid chromatography, possible.

REFERENCES

1. Huber, J.F.K., Hulsman, J.A.R. and Mejers, C.A.M. (1971). *J. Chromatog.* 62, 79-91.
2. Siggia, S. and Dishman, R.A. (1970). *Analyt. Chem.* 42, 1223-1229.

SEPARATION OF CORTICOSTEROIDS FOR MEASUREMENT BY COMPETITIVE PROTEIN BINDING

Joan Butler, Vera Fantl and C. K. Lim

*Department of Chemical Pathology,
King's College Hospital Medical School,
London.*

Congenital adrenal hyperplasia is a disorder of biosynthesis of the steroid hormone cortisol in the adrenal gland. It is inherited as a recessive condition from apparently normal parents. In severe cases the deficiency of cortisol production is life-threatening; in the first weeks of life the infant goes into a potentially fatal salt-losing crisis all too often mistaken for gastro-enteritis. A rapid diagnosis is therefore imperative even during steroid therapy.

The biosynthesis of cortisol and related steroids is shown below in a simplified form:

$$\text{precursors} \begin{cases} 17\alpha\text{-hydroxyprogesterone} \xrightarrow{\text{21-hydroxylase}} 11\text{-deoxycortisol} \xrightarrow{\text{11-hydroxylase}} \text{cortisol} \\ 11\text{-deoxycorticosterone} \longrightarrow \text{corticosterone} \end{cases}$$

The most common form of congenital adrenal hyperplasia is a deficiency of the enzyme 21-hydroxylase, leading to an accumulation of 17α-hydroxyprogesterone. To diagnose this disorder by demonstration of raised levels of 17α-hydroxyprogesterone in the blood, it is necessary to isolate this steroid from 11-deoxycortisol, cortisol, 11-deoxycorticosterone and corticosterone.

For this purpose it is necessary to be able to measure amounts of 1-10 ng of steroid in plasma samples of 100-200 µl, well below the sensitivity of ultraviolet absorption detectors. The method presented here uses high performance liquid chromatography for separation of the corticosteroids

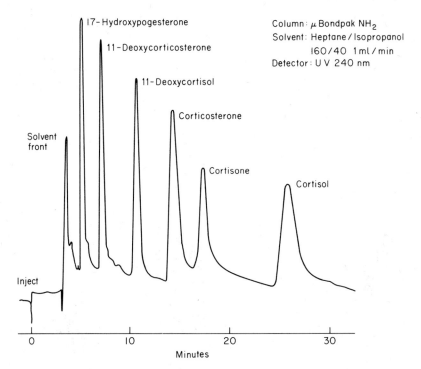

Fig. 1. Separation of Corticosteroids.

in a crude dichloromethane extract of plasma, and the competitive protein binding method for measurement of 17α-hydroxyprogesterone in fractions of the column eluate.

Separation on a μ Bondapak C_{18} column with the reverse-phase system methanol:water 50:50 gives poor resolution of 11-deoxycortisol and corticosterone. The eluate from this column is unsuitable for direct analysis by the competitive protein binding method which is best carried out on dried residues, and the retention time of 17α-hydroxyprogesterone is inconveniently long.

On the μ Bondapak $-NH_2$ column using normal phase solvent mixture of heptane:isopropanol 160:40, good resolution of all six steroids was obtained. (Fig. 1).

The order of elution is the same as that obtained with Sephadex LH 20 in cyclohexane:ethanol mixtures under gravity flow.

For estimation of 17α-hydroxyprogesterone one minute fractions were collected from this column, using the appearance of the solvent front on the recorder as zero time. Each fraction was evaporated to dryness and estimated by competitive protein binding. The recovery of tritium-labelled and unlabelled steroid was about 75% and the blanks about 1 ng. It is anticipated that a result could be obtained in under two hours.

It is hoped to extend this method to include measurement of 11-deoxycortisol and corticosterone for the detection of other disorders of corticosteroid biosynthesis.

SEPARATION OF PORPHYRIN ISOMERS BY HPLC - BIOCHEMICAL AND BIOSYNTHETIC APPLICATIONS

Alan R. Battersby, Dennis G. Buckley, Gordon L. Hodgson,
Roger E. Markwell and Edward McDonald

*University Chemical Laboratory,
Lensfield Road, Cambridge.*

Much is now known of the biosynthesis of porphyrins, chlorins and corrins, (1) although the details of certain parts of the biosynthetic pathway to these tetrapyrrolic pigments have not been determined. One area of current interest is the conversion of the monomeric porphobilinogen (PBG) specifically into uroporphyrinogen-III by the co-operative action of deaminase and cosynthetase, and a great variety of experimental studies have been made (1). The possible intermediacy of linear di-, tri-, and tetrapyrroles has been proposed, although unequivocal evidence for any specific intermediate between PBG and uroporphyrinogen-III has not been obtained. Active research in both Buenos Aires (2,4) and Cambridge (3,4) has centred on the possibility that one or more of the four possible aminomethylpyrromethanes (1)-(4) derived by condensation of two molecules of PBG is involved;

(1) = AP.AP; $R^1=R^3=A$, $R^2=R^4=P$
(2) = PA.AP; $R^1=R^4=P$, $R^2=R^3=A$
(3) = AP.PA; $R^1=R^4=A$, $R^2=R^3=P$
(4) = PA.PA; $R^1=R^3=P$, $R^2=R^4=A$

$A = CH_2CO_2H$, $P = CH_2CH_2CO_2H$

some aspects of these studies will be considered here.

All researches on the enzymic incorporation of PBG and aminomethylpyrromethanes into porphyrinogens are complicated by concomitant *non-enzymic* conversion of the substrates into porphyrinogens, and this results in a troublesome blank.
This is a major difficulty encountered by all workers in the area and undoubtedly it has contributed to reports of somewhat dissimilar results from different researchers. In order to obtain unequivocal evidence regarding the incorporatio of possible precursor aminomethylpyrromethanes (and tri- and tetrapyrroles), proof will be required (a) that the precursor is clinically (i.e. also isomerically) pure, (b) that the isolated porphyrin isomers, especially type-I and type-III, are chemically (i.e. also isomerically) pure, (c) that in the ^{14}C series, they are radiochemically pure, and (d) that the incorporations observed are specific and the site(s) of labelling are determined. The major difficulties in meeting some of the above criteria have now been overcome using new HPLC techniques, and the current position is described below.

The synthesis of the four isomeric aminomethylpyrromethanes (1)-(4) has been described (4) and is shown in outline for the PA.AP isomer (5) (Scheme 1); the synthesis of the other three isomers was accomplished analogously. Although the synthetic route used should produce isomerically homogeneous aminomethylpyrromethane precursors, it is essential to have independent proof that small quantities of the other isomers are not present in the material used for incubation studies due to the low values of the incorporations* of radioactivity (0.6-0.9%) usually encountered.

A mixture of the four isomeric aminomethylpyrromethane lactam benzyl esters (*cf*. (5), R = CO_2CH_2Ph), prepared by mixing the four separately synthesised isomers, was completely resolved on a Waters' 60 cm x 4 mm 10 µ Porasil column using dioxan-heptane (1:1) as eluting solvent (flow rate 2 ml/min; UV detector at 280 nm). The order of elution of these isomers (with retention volumes) is as follows: (i) PA.AP (77 ml), (ii) AP.AP (88 ml), (iii) PA.PA (98 ml), and (iv) AP.PA (121 ml). The identity of each peak on the chromato-

Incorporation values are calculated as percentages thus:

$$Incorporation = \frac{Total\ activity\ in\ isolated\ porphyrin}{Total\ activity\ in\ precursor\ used} \times 100\%$$

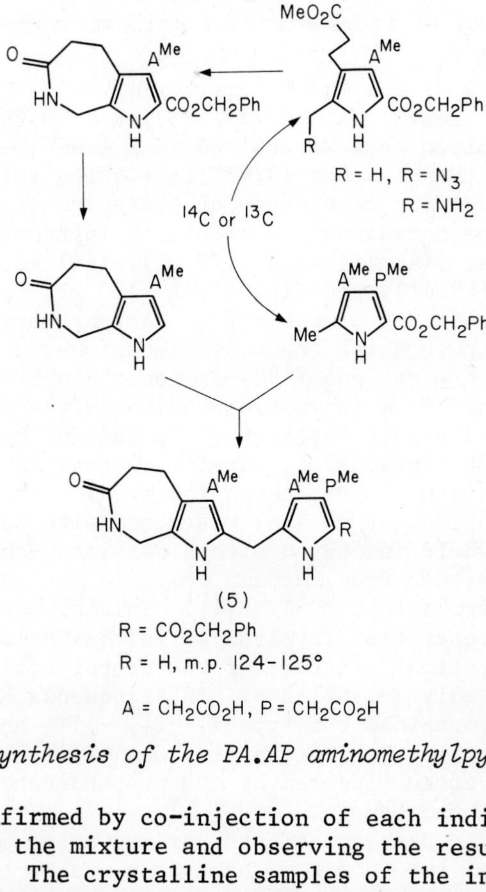

Scheme 1. *Synthesis of the PA.AP aminomethylpyrromethane isomer.*

gram was confirmed by co-injection of each individual isomer in turn with the mixture and observing the resultant peak enhancement. The crystalline samples of the individual aminomethylpyrromethane lactam benzyl esters were shown to be homogeneous by HPLC under the above conditions, and in particular were shown to be isomerically pure.

Removal of the benzyloxycarbonyl residue from each isomer separately was accomplished by hydrogenolysis over palladium black in methanol followed by decarboxylation of the carboxylic acid with cold anhydrous trifluoroacetic acid (4). The decarboxylated material was purified by p.l.c. on silica and crystallised from methanol to give the aminomethylpyrromethane α-free lactams (*cf.* (5), R = H) in good yield. Since acid treatment of the α-free lactams could have caused some rearrangement of the α-free pyrrole ring {e.g. (5), (PA.AP)→ (5) + isomer (PA.PA)}, it was also necessary to prove the isomeric purity of the individual aminomethylpyrromethane α-free lactams.

A mixture of the four isomers, (*cf.* (5), R=H), obtained from the pure benzyl esters (*cf.* (5), R = CO_2CH_2Ph) as above, was well resolved on a Waters' 60 cm x 4 mm 10 μ Porasil column using methanol-ether (3:47) as eluting solvent (2 ml/min; 240 nm). The order of elution of these isomers is identical to that of the corresponding aminomethylpyrromethane lactam benzyl esters, *viz*: (i) PA.AP (77 ml), (ii) AP.AP (85 ml), (iii) PA.PA (99 ml), and (iv) AP.PA (120 ml). The crystalline samples of each of the α-free lactams were shown to be isomerically pure using the above technique.

Each of the various (^{14}C)-aminomethylpyrromethanes (e.g. 2) was obtained from the corresponding α-free lactam (e.g. 5; R=H) by hydrolysis (4) with 2-$\underline{M}$ potassium hydroxide at 20-25° for 72h, followed by careful neutralisation using IRC-50 resin (H^+ form). For incubation studies the freshly neutralised solution (at pH 7.6) was mixed with radioinactive PBG and the whole incubated with a deaminase-cosynthetase preparation either from *Euglena gracilis* (at pH 7.6) (5) or from wheat germ (at pH 7.5) (2,4). Initial work with the wheat germ preparation involved aerial oxidation and dilution with a "statistical" mixture of the uroporphyrins I-IV (prepared by the polymerisation of PBG in aqueous hydrochloric acid) which contained the type-I, -II, -III, and -IV isomers in the ratio 1:1:4:2. The work-up was troublesome, and the analysis was often hindered by the preponderance of the type-III isomer in the sample.

Ideally, separation of the derived uroporphyrin esters was desirable, but so far we have been unable to separate any of these directly by HPLC. However, decarboxylation of uroporphyrins to coproporphyrins proceeds readily without rearrangement (6a), and since various t.l.c. techniques were known (6b) for separating most of the possible combinations of coproporphyrin isomers, we decided to examine the separation of the four coproporphyrin isomers by HPLC.

Initial results with 1.2 m/2 mm Corasil II columns under a variety of conditions were encouraging, but good separations could be achieved only in certain cases. Furthermore, the ultimate necessity for small scale (50-100 μg) preparative separations made it essential that columns of much higher capacity as well as increased resolving power should be used. The Waters 10 μ Porasil columns proved to be the best of those available for our needs; using two 30 cm x 4 mm 10 µ Porasil columns with ether-heptane (3:2) or dioxan-heptane (1:4) as eluting solvent, excellent separations of I + II, I + III and I + IV pairs of coproporphyrin isomers were achieved with flow rates in the range 0.5-2 ml/min. Under these conditions, a mixture of coproporphyrins III and IV tetramethyl esters could be separated for the first time, after 3-4 recycles (Waters' M6000 Pump) using ether-heptane (3:2) as eluting solvent (2 ml/min; 380 nm) (separation factor, α = 1.04 after 4 recycles).

Two special problems were encountered which precluded routine use of the above conditions for separating mixtures of all four coproporphyrins on 10 µ Porasil columns. The first was the observation that the resolution of coproporphyrins III and IV was reduced greatly by the presence of even small amounts of either of the other isomers, possibly due to association (6b) of the porphyrins in solution or on the surface of the silica. The other difficulty was <u>the slow deterioration of the resolving power of the columns *towards coproporphyrins*</u>; when standard resolution tests were carried out (with either a mixture of isomeric xylenols or nitrobenzene and methyl benzoate) the resolving power of the columns was shown to have remained close to the values obtained when the columns were new. It was hoped that these difficulties could be avoided by using reverse phase columns, and this was the case with the notable exception that the coproporphyrins III and IV could not be separated under any conditions we used. However, a mixture of coproporphyrins can now be partially resolved reliably and reproducibly into three fractions (I, II, III/IV) by using reverse phase HPLC, and the separation of the two unresolved isomers (III + IV) is possible on 10 µ Porasil columns using the recycle technique. The routine operations are described below.

The mixture of coproporphyrins obtained by decarboxylation of the diluted uroporphyrins from the enzyme incubation was converted to the mixture of coproporphyrin tetra-ethyl esters with 5% sulphuric acid in absolute ethanol. HPLC separation on two Waters' 30 cm x 4 mm 10 µ Bondapak/C_{18} reverse phase columns using acetonitrile-water (7:3) as eluting solvent (3 ml/min; 380 nm) gave three completely resolved fractions after 2 recycles (Fig. 1); coinjection of each of the four pure isomers showed that the isomers were eluted in the order I, III/IV (unresolved), and II. The unresolved III/IV mixture was transesterified to the mixture of tetramethyl

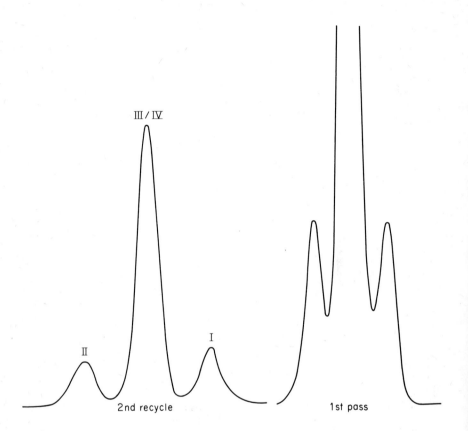

Fig. 1.

Sample:	*Coproporphyrins I-IV tetraethyl esters (ca. 1 μg)*
Column:	*2 x 30 cm x 4 mm 10 μ Porasil/C_{18} (Waters)*
Pump:	*M6000 (with UK6 Injector)*
Detector:	*Cecil C212 UV detector, 380 nm*
Solvent:	*Acetonitrile-water (7:3)*
Flow Rate:	*3 ml/min*
Chart Speed:	*2 min/cm*

esters and this was separated as before on 10 μ Porasil columns to give the two isomers in the order III then IV after 10-12 recycles (Fig. 2).

The above quantitative methods are being applied in our current work on the pyrromethanes (1)-(4) in relation to por-

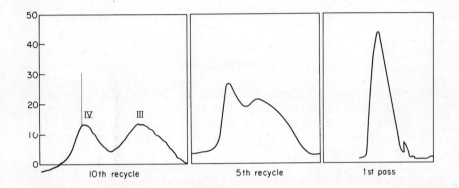

Fig. 2.

Sample:	*Coproporphyrins III + IV tetramethyl esters (ca. 10 µg)*
Column:	*2 x 30 cm x 4 mm 10 µ Porasil (Waters)*
Pump:	*M6000 (with septum Injector)*
Detector:	*Cecil C212 UV detector, 380 nm*
Solvent:	*Ether-heptane (2:3), 90% saturated with water*
Flow Rate:	*2 ml/min*
Chart Speed:	*5 min/cm*

phyrin biosynthesis and both enzymic and chemical aspects are being studied.

One result of this work which will be widely useful is the ready separation of coproporphyrins -I and -III, using either the normal or reverse phase columns, and of these methods we find the reverse phase technique to be the most reliable. This separation technique has proved to be very valuable in our studies of the various deaminase-cosynthetase enzyme systems, and will no doubt be very useful in clinical laboratories.

Acknowledgements
We give special thanks to Mr. G. R. Purdy for skilled assis-

tance with the various HPLC operations and to Drs. J. Saunders and D. C. Williams for help with the enzyme preparations and for valuable discussions. We thank St. Catherine's College, Cambridge for a Research Fellowship (to D. G. Buckley) and the S.R.C. and the Nuffield Foundation for financial support.

REFERENCES

1. Battersby, A.R. and McDonald, E. (1975). *In* "Porphyrins and Metalloporphyrins", 2nd Edn. Ch. 3, (K.M. Smith, ed.) Elsevier, Amsterdam and London.
2. Frydman, R.B., Valasinas, A. and Frydman, B. (1973). *Biochemistry*, 11, 80.
3. Battersby, A.R. (1971). 23rd International Congress of Pure and Applied Chemistry, Special Lectures, 5, 1.
4. References 90-93, 119, and 120 in Reference 1.
5. Carell, E.F. and Kahn, J.S. (1964). *Arch. Biochem. Biophys.* 108, 1.
6. Falk, J.E. (1964). *In* "Porphyrins and Metalloporphyrins", (a) p. 147; (b) Ch. 14, pp. 189-210. Elsevier, Amsterdam, London and New York.

HPLC ANALYSIS OF PORPHYRINS IN BIOLOGICAL MATERIALS

N. Evans, A. H. Jackson, S. A. Matlin and R. Towill

*Department of Chemistry,
University College, Cardiff.*

The main objectives of our work on the HPLC of porphyrins are to develop rapid quantitative techniques for the estimation of porphyrins formed either in the course of biosynthetic experiments, or as a consequence of disorders of metabolism, i.e. porphyrias (whether of genetic origin or caused by poisoning). We have also been interested in the use of HPLC for the isolation of small quantities of porphyrins for structure determination.

In most of our work it has been convenient to convert the porphyrin free acids, after isolation, to the corresponding methyl esters owing to the much greater solubility of the latter in organic solvents, not to mention the availability of a considerable amount of information on their behaviour on open column chromatography. The intense absorption maximum exhibited by porphyrins in the near U.V. at about 400 nm (the Soret band) provides a very convenient and sensitive means for their detection, and we have used a variable wavelength detector throughout our studies. Whilst a number of different types of column packing materials have been investigated including reversed-phase, most of the results described in this paper have been obtained with silica columns, either of the pellicular type (e.g. Corasil II) or microparticulate (e.g. Merckosorb SI 60).

Initially we used isocratic elution techniques with a variety of solvents and obtained good separations of porphyrin esters. The retention times were found to increase as the number of ester side-chains increased, but the later peaks were considerably broadened. This caused a loss in effective sensitivity, which could be obviated to some extent by increasing the flow rate through the column, and thus shortening the retention times. For the separation of two or three porphyrins differing only by one or two ester groups, however, isocratic elution has proved to be a very satisfactory method; moreover, it is very convenient experimentally if a number of similar determinations are to be carried out in sequence as the column does not need re-conditioning between runs.

For the separation of more complex mixtures of porphyrins we turned to gradient elution techniques, and for cheapness and simplicity the solvent gradient was formed at low pressure before being pumped onto the column. A conical glass mixing vessel was fitted with three side-arms, two of which were connected to solvent reservoirs (A and B) via glass taps, whilst the third was connected to the pump. Initially one solvent (A) was admitted to the flask, then the tap to this reservoir was closed, and the tap to the other (B) was then immediately opened. The concentration of the second solvent (B) then increases exponentially with time as the pump draws off the (well-stirred) mixture from the flask. Analyses of the effluent solvent were in good agreement with the calculated values, and it was relatively easy to change the solvent gradient, e.g. by altering the composition of the solvent in reservoir B.

In conventional open column chromatography of porphyrin esters chloroform or methylene chloride either alone or mixed with benzene and other hydrocarbons are very common eluents, as the esters are usually very soluble in the chlorinated solvents. However, solubility is not a major consideration for *analytical* HPLC and moreover chloroform can give rise to traces of acid which can be deleterious to metal columns, connecting tubing etc. Furthermore any metal ions produced may complex with the porphyrins. The chloroform is usually stabilised by the addition of small amounts of ethanol, but batch to batch variations may occur if commercial solvents are used directly; this can only be obviated by removing the ethanol completely, distilling the solvent, and then adding a known amount to the chloroform. For these reasons in most of our work we have used solvent systems such as hexane/ethyl acetate, or light petroleum (b.p. 60-80°)/ethyl acetate.

Figure 1 shows the separation of a mixture of porphyrins obtained from a human symptomatic porphyric urine and containing four to eight carboxyl groups; rapid and complete resolution was obtained within fifteen minutes or so of injection, and with full sensitivity for each component. Similar separations were obtained with standard mixtures of porphyrin esters. In the case of the natural mixtures confirmation

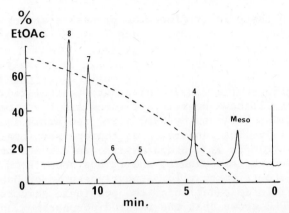

Fig. 1. Gradient Elution (hexane to ethyl acetate) of a mixture of Porphyrins obtained from the Urine of a Symptomatic Porphyric Patient on Merckosorb SI 60. (Mesoporphyrin-IX dimenthyl ester has been added as a standard).

of the identity of the individual fractions was provided by the relatively new technique of field desorption mass spectrometry. In this technique a special carbon fibre coated emitter wire is dipped into each of the eluates in turn before insertion into the mass spectrometer; essentially only molecular ions are obtained with this 'soft' ionisation method, and it is particularly suited to relatively involatile materials like porphyrins. Direct determination of the field desorption mass spectrum of the original mixture provides an additional qualitative check on the composition deduced from HPLC because only the molecular ions of each species present are observed. (1).

Whilst much of our work has been of an essentially analytical character we have also made good use of HPLC in separating relatively large amounts of porphyrins on a preparative scale. Perhaps the most fruitful example has been the separation of the components of a mixture of porphyrins obtained from the faeces of poisoned rats. Some 10-15 mg quantities of the octa-. hepta-, hexa-, penta-, and tetra-carboxylate fractions were separated on a column (1 in x 4 ft) of Porasil A-60 by gradient elution. Spectroscopic analyses and comparison with appropriate samples has enabled characterisation of each of these fractions. These results together with related biosynthetic studies have demonstrated that the conversion of uroporphyrinogen-III to coproporphyrinogen-III occurs by an essentially clockwise process starting with the decarboxylation of the D-ring acetic acid residue followed by those on the A, B and C rings (2).

Studies of porphyrin biosynthesis in haemolysates of red blood cells have also been greatly facilitated by HPLC. Incorporations of a variety of substrates related to uroporphyrinogen-III have been investigated and the time course of the reactions followed by HPLC after isolating the porphyrins produced (3). The conversion of coproporphyrinogen-III to protoporphyrin-IX by coproporphyrinogen oxidase has also been followed in this manner, and it has been shown that the intermediate tricarboxylic porphyrinogen is probably derived by a specific degradation of the propionic acid residue on ring A; HPLC shows that the related porphyrin (isolated in these experiments) is essentially a single isomer as it can be separated from the analogue in which the B ring propionic acid has been replaced by vinyl (4,5). (Their retention times differ by some three minutes under carefully controlled isocratic conditions, and baseline separation can be achieved). As a result of these and other experiments, which have depended very heavily on HPLC we have thus been able to define most of the biosynthetic stages between uroporphyrinogen-III and haem.

Another important aspect of our work on the HPLC of porphyrins has been the study of porphyrin profiles in porphyric urine and faeces (Fig. 1). For this type of investigation we needed to know not only the relative amounts of each porphyrin present but also the absolute amounts, and thus it was necessary to add a known amount of a standard compound. Mesoporphyrin-IX dimethyl ester was chosen for this purpose, in view of its stability and ready availability, and the fact that neither it nor protoporphyrin-IX (from which it is derived) normally occur even in porphyric urines; moreover it is eluted in advance of all the other naturally occurring porphyrins as it is a dicarboxylic material. Preliminary studies with various synthetic mixtures of porphyrins with 4, 5, 6, 7 and 8 carboxylic ester groups and mesoporphyrin-IX dimethyl ester confirmed that straight line calibration graphs of mole ratio porphyrin/mesoporphyrin versus area ratio porphyrin/mesoporphyrin could be obtained.

Porphyric urine samples were extracted by two techniques; (i) solvent extraction (6), or (ii) adsorption onto talc (7). The first procedure involved extraction of the urine (adjusted to pH 3) with ethyl acetate/butanol (1:1) followed by

TABLE I

Relative Percentages of Porphyrins in Porphyric Urines Estimated by Solvent Extraction and HPLC

No. of Carboxyl Groups	Patient			
	A	B	C	D
8	65	57	11	57
7	27	31	3	36
6	3	2	1	4
5	2	4	9	2
4	3	6	76	1

TABLE II

Relative Percentages of Porphyrins in Porphyric Urines Estimated by Adsorption on to Talc and HPLC

No. of Carboxyl Groups	Patient			
	A	B	C	D
8	54	45	28	74
7	24	33	8	20
6	6	4	4	3
5	3	7	9	1
4	13	11	51	2

further portions of butanol until neither phase showed any fluorescence under UV light. The combined organic extracts were then evaporated to dryness and the residue esterified with 5% sulphuric acid/methanol. The porphyrins adsorbed on talc were either eluted with HCl-acetone and then esterified, or eluted and esterified directly with sulphuric acid/methanol.

Preliminary results obtained from urine samples of five symptomatic porphyric patients are shown in Tables I and II. It can readily be seen that one of the patients has a significantly different excretion pattern from the other four, with less uroporphyrin and much more coproporphyrin. In the three examples where both extraction procedures were used there is a strong qualitative resemblance, but there are significant quantitative differences in the profiles. Further work is in progress on both methods but the reproducibility of the solvent extraction method seems better, although the talc method has some practical advantages.

The analysis of porphyrins in much smaller quantities (down to the pico-mole level) is also in progress, but although the sensitivity of HPLC is quite adequate metal complex formation can cause problems. In part the difficulties may have arisen because of the use of chloroform to extract the esters (after the esterification step), and Figure 2 shows the result of keeping one such dilute chloroform solution; after three days in the dark the meso- and copro-porphyrin esters are largely converted to their copper complexes in the HPLC system. This occurred in spite of the use of rigorously purified solvents and may be due to dissolution of traces of copper in the apparatus by acid formed in the chloroform. We have also observed the formation of zinc complexes in a case where a urine sample was inadvertently allowed to come into contact with a brass bolt. In the light of these results we have some doubts about the reported presence of zinc and copper porphyrins in some urines.

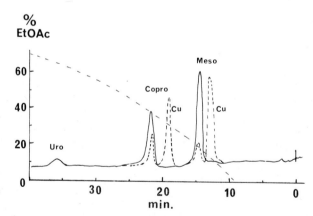

Fig. 2. Partial Transformation of a Chloroform Solution of Porphyrin esters into their Copper Complexes on HPLC. (a) — run immediately, (b) ... after 3 days standing.
(Total weight of porphyrin ~ 350 µg in 10 ml $CHCl_3$ containing 2% EtOH)

Further studies of both porphyric and normal urines are in progress to determine the best procedure both for the extraction and for HPLC. There is much to be said for the direct analysis of porphyrin free acids, and preliminary results with an ion-exchange column look promising, but much development work needs to be done. The potentially greater sensitivity of fluorescence detection in the monitoring of porphyrin free acids offers the possibility of analysing trace amounts in extracts from specific organs or tissues; much of the improvement is lost in studying the esters owing to quenching by organic solvents.

All of the work reported in this paper was carried out with a Waters Model 6000 pump and septum injector, using stainless steel columns, and a Cecil variable wavelength detector fitted with a 10 µl flow cell.

Acknowledgements
We thank the Medical Research Council and the Royal Society for generous financial support and we are grateful to Dr. D. Dean (Liverpool), Dr. G. H. Elder and Dr. S. G. Smith (Cardiff) for kindly providing the porphyric materials.

REFERENCES

1. Evans, N., Games, D.E., Jackson, A.H. and Matlin, S.A. *J. Chromatogr.* In press.
2. Jackson, A.H., Sancovich, H.A., Ferramola, A.M., Evans, N., Games, D.E., Matlin, S.A., Elder, G.H. and Smith, S.G. *Phil. Trans. Roy. Soc.* (London). In press.
3. Jackson, A.H., Ferramola, A.M., Sancovich, H.A., Evans, N., Matlin, S.A., Ryder, D.J. and Smith, S.G. *Clin. Chim. Acta.* In press; Smith, S.G., Ferramola, A.M., Sancovich, H.A., Evans, N., Matlin, S.A., Ryder, D.J. and Jackson, A.H. *Clin. Chim. Acta.* In press.
4. Games, D.E., Jackson, A.H., Jackson, J.R., Belcher, R.V. and Smith, S.G. Unpublished work. (see also Ref. 1).
5. Cavaleiro, J.A.S., Kenner, G.W. and Smith, K.M. *J. Chem. Soc. Chem. Comm.* 1973, 183 and *J. Chem. Soc. Perkin I*, 1974, 1188.
6. Fernandez, A.A., Henry, R.J. and Goldenberg, H. *Clin. Chem.* 1966, $\underline{12}$, 463.
7. Falk, J.E., Porphyrins and Metablloporphyrins, Elsevier, Amsterdam, 1964, p.169.

PORPHYRIN PROFILES IN PORPHYRIAS

C. H. Gray, C. K. Lim and D. C. Nicholson

*King's College Hospital Medical School,
London.*

Diagnosis of diseases of porphyrin metabolism necessitates the routine identification and determination of porphyrins in blood, urine and faeces. Classical methods for such analyses utilised the hydrophilic and basic properties of porphyrins whereby those with less than five carboxyl groups were extracted into ether while those with five to eight carboxyl groups were extracted, at appropriate pH, into other solvents. The individual porphyrins were then determined spectrophotometrically or fluorimetrically in dilute aqueous acid solutions (1). Such fractions are now known, from studies by column-(2), paper-(3) and thin layer chromatography (4), to be complex and some twelve porphyrins are now considered important for the diagnosis of the porphyrias (5).

Table I shows the well known sequence of reactions which lead to the formation of haem and indicates those regions of the pathway affected in each of the known porphyrias. The porphyrins of greatest diagnostic value are the octacarboxylic uroporphyrins, heptacarboxylic porphyrins, the tetracarboxylic coproporphyrins and the dicarboxylic proto- and mesoporphyrins. It is necessary to consider also an isomer of coproporphyrin, namely isocoproporphyrin, shown by Elder (6, 7) to be important in the diagnosis of symptomatic porphyria.

Urinary porphyrins alone are not definitive with respect to the porphyrias but consideration of total urinary porphyrins and the pattern of the faecal porphyrins (Table II) allows easy differentiation of these diseases. We now wish to report the patterns of urinary and faecal porphyrins found by HPLC in variegate-. symptomatic-, hereditary copro-, erythrohepatic and congenital-porphyria and we submit that this technique provides a rapid and effective routine method for identifying these conditions. The results of these studies are shown in Table II.

TABLE I

Regions of the Haem Biosynthetic Pathway Affected in the Porphyrias

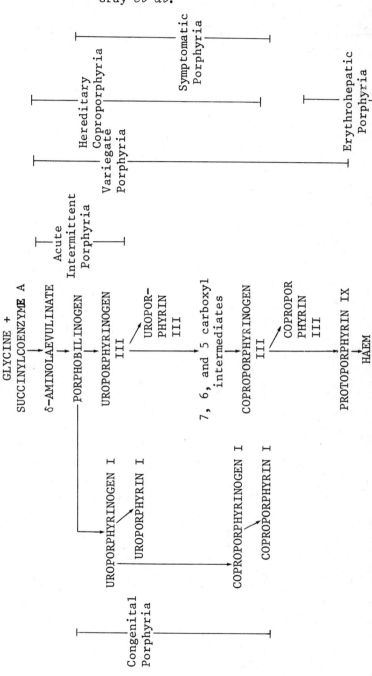

TABLE II

Total Urinary Porphyrins and Faecal Porphyrins Excreted in the Porphyrias

	Total Urinary Porphyrins	Faecal Porphyrins			
		Hepta	Isocopro	Copro	Dicarb-oxyl
Symptomatic Porphyria	++++	+++	++++	+ or normal	+ or normal
Variegate Porphyria	++ or normal	normal	normal	+++	++++
Hereditary Coproporphyria	+++	normal	normal	++++	++ or normal
Congenital Porphyria	++++	normal	normal	+++	++ to normal
Erythrohepatic Porphyria	normal*	normal*	normal*	normal*	++++

*When cirrhosis is present in erythrohepatic porphyria the urine may contain excessive quantities of porphyrins.

For this work porphyrins were isolated in a semi-pure state by classical method and chromatographed as methyl esters or were chromatographed from crude mixtures of total esterified porphyrins from urine and faeces. Details of the HPLC system used are included in Figures 1 and 2.

The differentiation of variegate and symptomatic porphyrias is important for clinical reasons. Characteristically a considerable excess of protoporphyrin is apparent in the faeces of the variegate patient and is accompanied by coproporphyrin which is also abnormally high. By contrast, the faecal pattern for symptomatic porphyria shows an excessive amount of a tetracarboxylic porphyrin readily distinguished from coproporphyrin by its different retention time and which is identical with that of Elder's isocoproporphyrin; the same chromatogram shows the presence of some coproporphyrin (Fig. 1).

Study of urinary porphyrins in variegate porphyria is

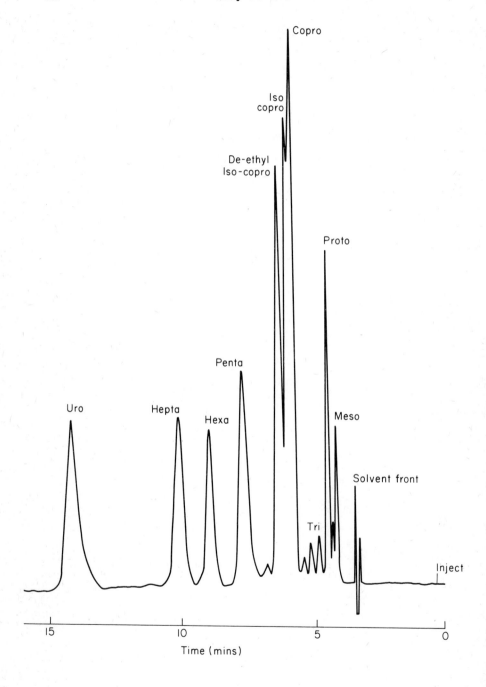

Fig. 1. Symptomatic Porphyria (PCT) Faecal Porphyrins.
Pump: Waters 6000
Column: µ Porasil 30 cm x 4 mm
Solvent: n-Heptane 3, Methyl Acetate 2
Flow Rate: Program
Detection: UV 404 nm, Cecil CE 212

useful only during acute exacerbation of the disease and our patient was in clinical remission. However, porphyrins in the urine of the patient with symptomatic porphyria provided peaks corresponding to excessive uroporphyrin and heptacarboxylic porphyrin. As with the corresponding faecal porphyrins, isocoproporphyrin and a little coproporphyrin were also present in urine although not excessively. Further differentiation between variegate and symptomatic porphyria depends on the fact that urinary excretion of the precursors porphobilinogen and aminolaevulinic acid during acute attacks of the former condition is never seen in the latter.

Hereditary coproporphyria, like variegate porphyria is an inherited disease usually asymptomatic, but sometimes with neurological manifestations. It has long been known to cause an overproduction of coproporphyrin. This is confirmed by the presence of large peaks for both urinary and faecal coproporphyrins and the HPLC shows complete absence of isocoproporphyrin from the excreta of a patient with this disease. Reverse phase HPLC on two µ Bondapak C_{18} columns with 70% aqueous acetonitrile was used to define the isomer type composition of the coproporphyrin in hereditary coproporphyria and was shown to be 95% isomer III.

Erythrohepatic protoporphyria is a photosensitising condition characterised by excessive faecal and blood porphyrins. The urine is completely devoid of porphyrins except when liver cirrhosis complicates late stages of the disease. Our first study of this condition was that of the faecal porphyrins from a young patient (LM) whose liver function was otherwise normal and showed an excessive excretion of protoporphyrin. For this patient there was no evidence of excessive urinary porphyrins and the faeces contained little uroporphyrin. By contrast the faeces from an elderly patient (HC) with a long history of porphyria, and who developed a severe and fatal cirrhosis, confirmed in addition to the expected protoporphyrin excessive uroporphyrin and small quantities of penta-, hexa-, and heptacarboxylic porphyrins (Fig. 2). The urine of this patient contained much copro- and uroporphyrin pre-

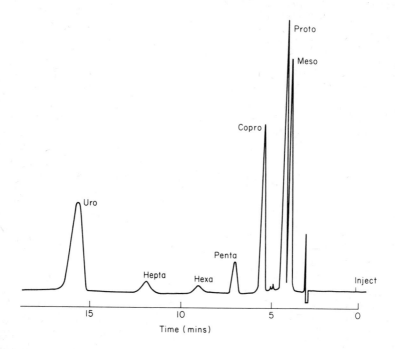

Fig. 2. Erythrohepatic Protoporphyria. Faecal Porphyrins: Cirrhosis.
Column: μ Porasil
Solvent: n-Heptane 3, Methyl Acetate 2
Flow Rate: 1.5 ml/min
Detection: UV 404 nm

sumably due to the hepatic abnormality superimposed on his primary disease. Since HPLC showed the absence of excessive heptacarboxyl- and isocopro-porphyrin this acquired condition must differ from symptomatic porphyria.

Finally, *congenital porphyria* is a rare condition in which an enzyme defect causes over production of Type I porphyrin isomers (Table I) especially coproporphyrin I and uroporphyrin I. The chromatograms of the faecal porphyrin from a patient (MD) with established congenital porphyria, shows little protoporphyrin and mesoporphyrin, considerable coproporphyrin and smaller quantities of hepta-, hexa- and pentacarboxyl porphyrins. The urine from this patient contains excessive uro- and coproporphyrins together with a moderate amount of pentacarboxylic porphyrin.

Congenital porphyria, with its very obvious clinical and haematological manifestations rarely requires biochemical differentiation from other conditions causing excretion of excessive uroporphyrin. Recently we have studied an adult (EA) who at the age of 54 years experienced sudden onset of photosensitivity. This patient was diagnosed as having symptomatic porphyria and, accordingly, had a high urinary excretion of uroporphyrin and coproporphyrin. The HPLC of the faecal porphyrins and urinary porphyrins, however, were identical with those of our well established congenital porphyric (MD) and the porphyrins were later found to be mainly Type I isomers. EA was thus a rare example of late onset congenital porphyria.

The characteristic chromatograms obtained for urine and faeces in the various types of porphyria and the rapidity and simplicity of the investigation makes HPLC admirably suitable for routine work.

Acknowledgement
We are grateful to Mr. W. Lockwood for helpful discussion and advice.

REFERENCES

1. Rimington, C. (1971). The Association of Clinical Pathologists Broadsheet.
2. Vannotti, A. (1954). "Porphyrins: Their Biological and Chemical Importance". Hilger and Watts, London.
3. Chu, T.C., Green, A.A. and Chu, E.J. (1951). *J. Biol. Chem.* 90, 643-646.
4. Doss, M., Meinhof, W., Look, D., Henning, D., Nawrock, P., Dolle, W., Strohmeyer, G. and Filippini, L. (1971). *South African J. Lab. Clin. Med.* 17, 50-54.
5. Gray, C.H., Elder, G.H. and Nicholson, D.C. (1972). *J. Clin. Path.* 75, 1013.
6. Elder, G.H. (1972). *Biochem. J.* 126, 877-891.
7. Elder, G.H. (1975). *J. Clin. Path.* 28, 601-607.

APPLICATION OF HPLC TO THE ANALYSIS OF CLINICALLY IMPORTANT PORPHYRINS

R. E. Carlson and D. Dolphin

*Department of Chemistry,
The University of British Columbia,
Vancouver.*

The porphyrias are a group of metabolic diseases characterized by the abnormal production and excretion of porphyrins. The clinical diagnosis of porphyria is based on the qualitative and quantitative determination of the pattern of excreted porphyrins which has been shown to be characteristic of the type of porphyria (1).

The analytical methods currently used for porphyrin analysis are difficult, laborious and expensive. We recently (2) developed a high pressure liquid chromatographic technique for the quantitative analysis of the urinary porphyrins. We have now studied the application of a high pressure liquid chromatographic method to the analysis of fecal porphyrins, which may include, besides the 2 to 8 carboxylated compounds, a heterogeneous group of sub-uroporphyrin compounds (3,4).

Experimental

Fecal Sample Extraction and Column Chromatography

20 ml of methanol and 2 ml of $BF_3 \cdot Et_2O$ were added to the fecal sample (.5 g wet or .2 g dry), and the solution was stirred overnight. The esterified material was extracted with two 100 ml portions of methylene dichloride after the addition of 80 ml of water. The methylene dichloride solution was dried over sodium sulfate, filtered and the solvent removed on a rotary evaporator.

The residue was dissolved in a minimum of methylene dichloride and chromatographed on 30 g of silica gel Woelm activity IV in a 1.6 cm x 45 cm glass column. The silica was packed in the column using methylene dichloride/methanol

(99:5:0.5 v/v), and was eluted with the same solvent to remove a yellow-brown material. Elution by methylene dichloride/ethyl acetate/methanol (45:15:10) gave the 2 to 8 carboxyl porphyrins followed by a band of sub-uro material. The eluent between the major porphyrin band and the sub-uro material was fractionally collected and compared to uroporphyrin on silica gel TLC (methylene dichloride/ethyl acetate/methanol, 75:30:5 v/v) to prevent any carboxyl porphyrin loss. The appropriate fractions were pooled and the solvent was removed on a rotary evaporator. The 2 to 8 carboxyl porphyrin residue was dissolved in methylene dichloride (1.0 ml/0.10 g feces) and analysed by HPLC.

Column Chromatography of the Sub-Uro Material

The sub-uro residue was dissolved in a minimum of methylene dichloride/methanol (98:2 v/v) and chromatographed on 10 g of silica gel Woelm activity IV which had been packed using the same solvent. After collection of the eluted band the solvent was removed on a rotary evaporator. The residue was dissolved in methylene dichloride (1.0 ml/0.10 g feces) and analysed by HPLC.

High Pressure Liquid Chromatography

Samples were analysed on a Waters Associates (WA) ALC 202 using ⅛ in x 2 ft column with Polyamide, Corasil C_{18}, Basic Alumina, Neutral Alumina, Porasil C or Corasil II packings (6). A Cary 17 spectrometer with a detection cell as described previously (2) was used for detection at 300 and 403.5 nm with a chart speed of 0.4 in/min and for visible scans (350-650 nm) taken during chromatographic analysis. The 254 nm chromatogram was recorded using a Waters Associates UV detector.

The gradient for sample elution was generated using a 12.5 ml loop in a Waters Associates valve and loop injector. The loop was made from 3.8 meters of 2 mm I.D. stainless steel tubing. The pump was constantly supplied with the higher polarity solvent and uses displacement of the lower polarity solvent from the loop for proper sample elution. The time from loop "injection" to sample injection is about 2.5 minutes.

Solvent Preparation

n-Propanol and triethylamine (TEA) were reagent grade. The light petroleum (30-60°) and methylene dichloride used for HPLC were glass distilled and contained 10 μl TEA/100 ml. The glass distilled methylene dichloride-TEA solution was used within 5 days of distillation. Older solutions exhibited altered chromatographic properties.

Analysis of the 2 to 8 Carboxyl Porphyrin Sample

The sample was eluted with light petroleum (30-60°) (TEA)/methylene dichloride (TEA) (40:100) followed by light petroleum (30-60°)-TEA/methylene dichloride-TEA/n-propanol (15:100:.50). The 40:100 solution was placed in the loop in preparation for each injection. Solvent flow was 1.0 ml/min (< 50 psi). The eluted porphyrins were characterized by comparison with authentic samples as described previously (2).

Analysis of the Sub-Uro Material

The sample was eluted with methylene dichloride-TEA/n-propanol (100:2) followed by methylene dichloride-TEA/n-propanol (90:10). The 100:2 solution was placed in the loop in preparation for each injection. Solvent flow was 1.0 ml/min (< 50 psi). The visible spectra from the chromatogram were obtained by stopping the solvent flow and scanning the desired wavelength range.

RESULTS AND DISCUSSION

Evaluation of Available Column Packings

Chromatography of Porphyrin Free Acids

The development of a chromatographic system using Polyamide was unsuccessful because a solvent combination which would elute Proto-porphyrin IX could not be found. A reverse phase (C_{18}) column was tested with combinations of:

        Methanol   2% Ammonium Hydroxide
        Acetonitrile H_2O

The samples eluted with tailing or as very broad peaks.

Chromatography of Porphyrin Methyl Esters

The adsorption packings were tested with combinations of:

| Methylene Dichloride | Methanol Propanol | Ethyl Acetate Ethyl Propionate Propyl Acetate Propyl Propionate | Light Petroleum (30-60°) | Triethylamine |

Basic alumina gave poor resolution with broad peaks, while Porasil C gave good resolution but extremely broad peaks. Samples chromatographed on neutral alumina tailed too much to give acceptable resolution.

Corasil II gave good resolution and peak width using a methylene dichloride/light petroleum (30-60°)/n-Propanol/Triethylamine (TEA) system. This packing was applicable to a simple gradient system. If an electronic programmer were available methylene dichloride/n-propanol (0.3-1.0% n-propanol) would probably give a good chromatogram. However, for routine clinical analysis the operationally simpler system was best.

A comparison of standard and urine samples on Corasil II and Porasil T (Fig. 1) demonstrates that Porasil T gives the better resolution in the 4 to 8 carboxyl region and is well suited to the analysis of urine samples in which the 4 to 8 carboxyl porphyrins is important. For the analysis of fecal samples in which the 2 to 4 carboxyl porphyrins dominate Corasil II is prefered but for routine clinical needs either column is equally suitable.

Generation of a Gradient

The use of a valve and loop injector provided a convenient and reproducible method for the generation of a gradient. The baseline at 254 nm (Fig. 2) illustrates the change between the two solvent systems. This is supported by the tight uroporphyrin peak in the reference chromatogram because this compound elutes as the propanol/amine is flushed from the column.

Fecal Analysis

Boron trifluoride/methanol has been shown (5) to be an efficient method for the esterification of fecal porphy-

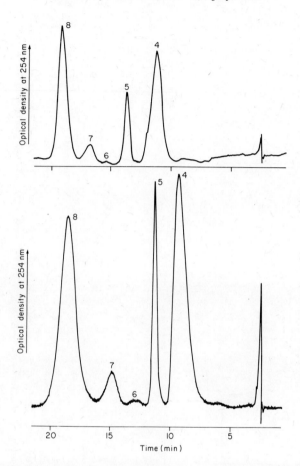

Fig. 1. Comparison of a urine sample chromatogram on Corasil II (upper) and Porasil T (lower). Note the difference in elution time for the 4 to 8 carboxylated porphyrins. Absorption not to scale.

rins. We combined this procedure with column chromatography on silica with methylene dichloride/ethyl acetate/methanol to give porphyrin samples which were suitable for HPLC analysis; this gave a convenient separation of the 2 to 8 carboxyl porphyrins from the sub-uroporphyrin (sub-uro) compounds The latter was rechromatographed on a methylene dichloride/ methanol column. With each fraction, fluorescence remained on the top of the column.

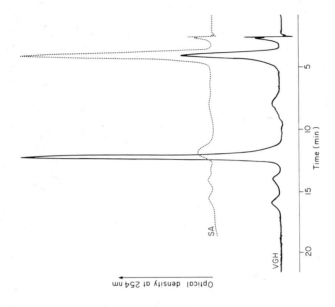

Fig. 3. Chromatogram of Fecal samples VGH and SA; absorption not to scale.

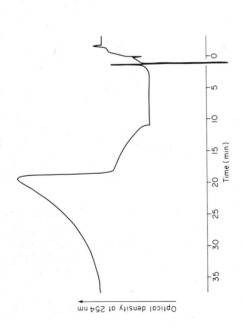

Fig. 2. Illustration of the solvent gradient (254 nm).
The spike at -2 minutes is caused by an air bubble and marks the change to lower polarity solvent. Methylene dichloride is injected at 0 minutes; the solvent front appears at 1.5 minutes, and the higher polarity solvent begins to elute at 11 minutes. The peak at 20 minutes is discussed in the text.

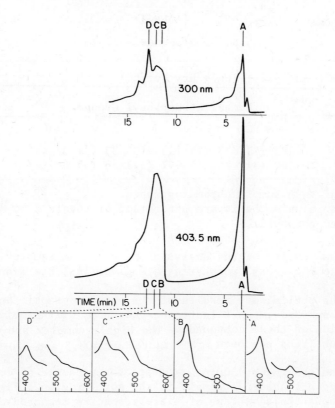

Fig. 4. Chromatogram and visible spectra of the sub-uro fraction.
The adsorption of the 300 and 403.5 nm chromatograms is on scale.
The visible spectra A-D were taken at the points indicated. The 400 nm region of A, C and D is reduced by a factor of 2.5.

TABLE I

Results of Feces Analysis

Porphyrin Analysis: (µg/g Dry Weight)

Sample	Number of Carboxyl Groups						
	2	3	4	5	6	7	8
S.A.	883 (79)	23 (2)	148 (13)	33 (3)	37 (3)	-	-
V.G.H.	540 (31)	73 (4)	997 (57)	56 (3)	66 (4)	-	13 (1)

50 µl of each sample chromatographed.
Figures in parenthesis are percentage of totals 2 to 8 carboxylated porphyrin.

The results of the analysis of two fecal samples for the 2 to 8 carboxyl porphyrins are given in Table I and Figure 3.

The preliminary results of the chromatographic analysis of the sub-uro group illustrates that HPLC will provide information on these compounds. The highly complex sub-uro fraction (VGH) has been investigated at 403.5 and 300 nm (Figure 4.

The tracing at 300 nm suggests the presence of compounds which do not absorb strongly in the soret region. Figure 4 also shows the advantage of visible spectra measured during a chromatographic scan. All four spectra (A-D) demonstrate the presence of porphyrins and spectrum C demonstrates the presence of a major non-Soret component.

Acknowledgement
This work was supported by the Canadian National Research Council and the United States National Institutes of Health (AM 17989).

REFERENCES

1. Chu, T.C. and Chu, E.T.H. (1967). *Clin. Chem.* <u>13</u>, 371-387.
2. Carlson, R.E. and Dolphin, D. (1975). *In* "First International Porphyrin Meeting: Porphyrins in Human Diseases". Freiburg. In press.
3. Rimington, C., Lockwood, W.M. and Belcher, R.N. (1968).

Clin. Sci. 35, 211-247.
4. Rimington, C. and With, T.K. (1974). *Enzyme*, 17, 17-28.
5. Grosser, Y. and Eales, L. (1973). *South African Med. J.* 47, 2162-2168.
6. The polyamide packing was a nylon type absorbent bonded to an impermeable bead available from Reeve Angel under the name Pellamidon. The Corasil C_{18} was a linear 18-carbon alkane bonded to a silica coating on an impermeable bead. It is sized at 37-50 µ and is available from Waters Associates. The basic and neutral Woelm aluminas were 18-30 µ, from Waters Associates. The Porasil C was a pherical totally porous silica of 37-75 µ size, from Waters Associates. The Corasil II was a silica coating on an impermeable bead. It was 37-50 µ from Waters Associates.

THE SEPARATION OF BILE PIGMENTS BY HPLC

M. S. Stoll*, C. K. Lim** and C. H. Gray**

*Bromley Accident Centre, Bromley Hospital,
Bromley, Kent.

**Department of Chemical Pathology,
King's College Hospital Medical School,
London.

Bile pigments are linear tetrapyrroles arising by oxidative fission of the porphyrin macrocycle with loss of a bridge carbon atom and introduction of terminal α oxygen atoms. The compounds are classified primarily according to the extent of the conjugation which varies from the green or blue coloured verdins with all four rings in conjugation to the colourless urobilinogens with no conjugation. Most natural bile pigments, derived from protoporphyrin IX by fission at the α bridge, have the IXα order of side chains, such as the important yellow pigment bilirubin IXα (i).

(i)

A large number of bile pigments belonging to six or more classes have been described, some are synthetic, others are naturally occuring from both animal and vegetable sources.

At present measurements of serum levels of bilirubin and its conjugates are of major clinical value and levels of urinary and faecal urobilin are of more limited value, but there is no obvious advantage in a knowledge of the detailed bile pigment profile in biological material comparable to that, for example, in the porphyrin field. However, it is known that the extent of reduction of bilirubin in the gut is dependent both on the nature of the intestinal flora and on the total output of bile pigment, and an ability to perform a detailed analysis routinely might reveal hitherto unsuspected phenomena.

Many methods for the separation of bile pigments have been devised including solvent partition, electrophoresis and chromatography. Recently the most successful separations have been achieved by thin layer chromatography (1,2,3). The high resolution offered by this technique led to the discovery of symmetrical IIIα and XIIIα compounds in preparations from purely IXα starting materials. This was subsequently shown to be due to acid-catalysed rearrangement of rubins (4) and urobilinogens and there is no evidence at present for the natural formation of the symmetrical isomers.

With the exception of rubins and urobilinogens all known classes of bile pigment are basic, basicity decreasing with increasing conjugation. This greatly influences separation, the order of mobility being urobilins<violins<verdins corresponding to two, three and four rings in conjugation, respectively.

The most generally useful t.l.c. system for verdins and violins dimethyl esters was carbon tetrachloride/dichloromethane/methyl acetate/methyl propionate on silica gel. The retention of urobilins was so great that a more polar system, less suitable for verdins and violins had to be used. This system is described later. In the present study no separations of the neutral rubin or urobilinogen dimethyl esters are described.

The t.l.c. system for verdins and violins was simplified for HPLC to carbon tetrachloride/methyl acetate, a system which has also been used to t.l.c. (5). Initial experiments were performed on a 3 ft x $\frac{1}{8}$ in column of Corasil II (Waters Associates), a pellicular packing of about 37 μ particle size. Since the only ultraviolet detector available was a

254 nm fixed wavelength instrument and since at this wavelength the solvent absorbed strongly and bile pigments absorbed poorly a refractive index detector had to be used. It was at once apparent that sensitivity with this detector was very low and results using this system were inferior to t.l.c. probably due to overloading of the Corasil II necessary to achieve adequate detection.

Use of a detector with a variable wavelength (Cecil Instruments) was a great advance making it possible to monitor at the ultraviolet absorption maximum of the compounds under study, or if different classes of pigment were simultaneously chromatographed a compromise wavelength could be chosen. There was now no solvent restriction since monitoring was always above 330 nm at which the solvents used had no appreciable absorption. Excellent sensitivity was achieved with this monitor, new solvent systems were developed and good separations, comparable with those obtained by t.l.c. were achieved on 6 ft x ⅜ in Corasil II in several solvent systems including methyl acetate or ethyl methyl ketone with cyclohexane or petroleum. Best results were obtained, however, with methyl acetate/isooctane which was used for all verdin and violin separations to be described. On Corasil II separations of geometrical isomers could only be achieved by using a high proportion of isooctane to methyl acetate to give a k' of between five and ten and by using a very slow flow rate, which due to relatively slow equilibration, substantially increased resolution. Analysis times of one to two hours were necessary with this system in the separation, for example, of mesobiliverdins IIIα, IXα and XIIIα dimethyl ester. A separation of eleven pigments in this system is shown in Figure 1a.

The introduction of µ Porasil (Waters Associates) a porous silica packing of 10 µ particle size, represented a great advance and with a 30 cm x ⅜ in commercially packed column excellent resolution superior to that achieved by t.l.c. could be obtained in times of 10 to 20 minutes. All subsequent separations to be described in this paper were performed on this column. Figure 1b shows separation of a similar mixture on µ Porasil as that shown in Figure 1a on Corasil II. The greatly increased resolution is apparent.

It has been found that both absolute and relative k' values depend critically on the solvent composition and the activity of the column packing and the values quoted in the tables are intended only as a guide to separation.

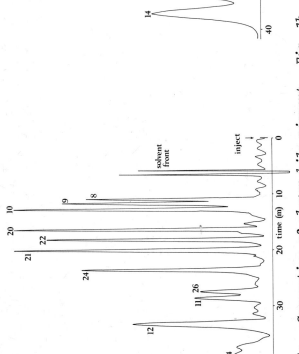

Fig. 1a. Separation of eleven bile pigment dimethyl esters on 6 ft x ⅜ in Corasil II in methyl acetate/isooctane (1/2) with detection at 330 nm. Flow rate programme. Peak numbers refer to compounds in Tables I, II and III.

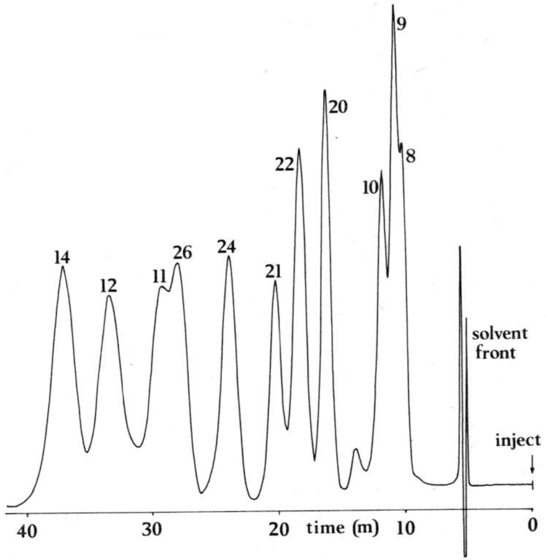

Fig. 1b. Separation of eleven bile pigment dimethyl esters on 30 cm x ⅜ in μ Porasil in methyl acetate/isooctane (2/3) with detection at 345 nm. Flow rate programme. Peak numbers refer to compounds in Tables I, II and III.

TABLE I

End Ring β Substituents and k' Values on μ Porasil in Methyl Acetate/Isooctane (1/2) of Ten Verdins (Structure ii)

Compound	R_1	R_2	R_3	R_4	k'
1	vinyl	methyl	methyl	vinyl	2.6
2	methyl	vinyl	methyl	vinyl	3.0
3	methyl	vinyl	vinyl	methyl	3.3
4	vinyl	methyl	methyl	ethyl	2.7
5	vinyl	methyl	ethyl	methyl	2.9
6	methyl	vinyl	methyl	ethyl	2.9
7	methyl	vinyl	ethyl	methyl	3.2
8	ethyl	methyl	methyl	ethyl	2.8
9	methyl	ethyl	methyl	ethyl	3.0
10	methyl	ethyl	ethyl	methyl	3.3

Table I shows the end ring β substituents and k' values of ten verdins in methyl acetate/isooctane (1/2) with detection at 360 to 380 nm. These compounds have the Structure ii.

(ii)

For comparison the k' values of mesobiliviolins in the same solvent system were about 15.

Table II shows the end ring β substituents and k' values of nine violins in methyl acetate/isooctane (1/2) with detection at 330 to 340 nm. These compounds have the Structure iii.

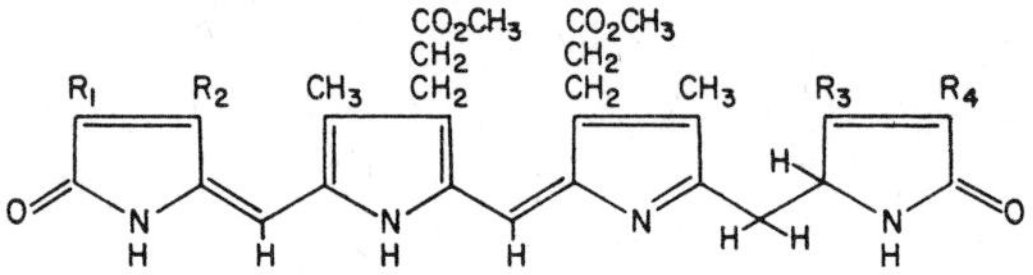

(iii)

TABLE II

End Ring β Substituents and k' Values on μ Porasil in Methyl Acetate/Isooctane (1/2) of Nine Violins (Structure iii)

Compound	R_1	R_2	R_3	R_4	k'
11	ethyl	methyl	methyl	ethyl	4.7
12	methyl	ethyl	methyl	ethyl	5.5
13	ethyl	methyl	ethyl	methyl	5.5
14	methyl	ethyl	ethyl	methyl	6.4
15	vinyl	methyl	methyl	ethyl	3.9
16	vinyl	methyl	ethyl	methyl	4.7
17	methyl	vinyl	methyl	ethyl	5.5
18	methyl	vinyl	ethyl	methyl	6.4
19	methyl	vinyl	methyl	vinyl	3.4

Table III shows the end ring β substituents and k' values of seven violins containing ethylidene substituents. Detection was at 330 to 340 nm and the solvent system was methyl acetate/isooctane (1/1). These compounds have the Structure iv.

(iv)

TABLE III

End Ring β Substituents and k' Values on μ Porasil in Methyl Acetate/Isooctane (1/1) of Seven Violins with Ethylidene Substituents (Structure iv)

Compound	R_1	R_2	Stereochemistry	k'
20	vinyl	methyl	Structure iv	2.2
21	methyl	vinyl	Structure iv	2.9
22	ethyl	methyl	Structure iv	2.5
23	methyl	ethyl	Structure iv	2.9
24	vinyl	methyl	Diastereoisomer of Structure iv	3.8
25	methyl	vinyl	Diastereoisomer of Structure iv	5.0
26	ethyl	methyl	Diastereoisomer of Structure iv	4.4

Since the inner α and the outer β carbon atoms of the unconjugated end ring are asymmetric these compounds exist in RR and RS forms. Although there is no evidence at present for the existence of these compounds in mammalian systems the compounds phycoerythrobilin and phycocyanobilin, found in red and green algae, are of closely related structure.

For urobilin dimethyl esters one of the best t.l.c. solvent systems was benzene/ethanol/ammonia (9/1/trace) (1). In continuous flow systems it was found that a less volatile base gave greater stability and consequently benzene/ethanol/ diethylamine was employed. This system gave excellent separations of isomeric urobilins with detection at their absorption maximum of 450 nm.

Table IV shows the end ring β substituents and k' values of six isomeric i-urobilins and Figure 2 shows their HPLC separation in benzene/ethanol/diethylamine (72500/2500/1). These compounds have the Structure v.

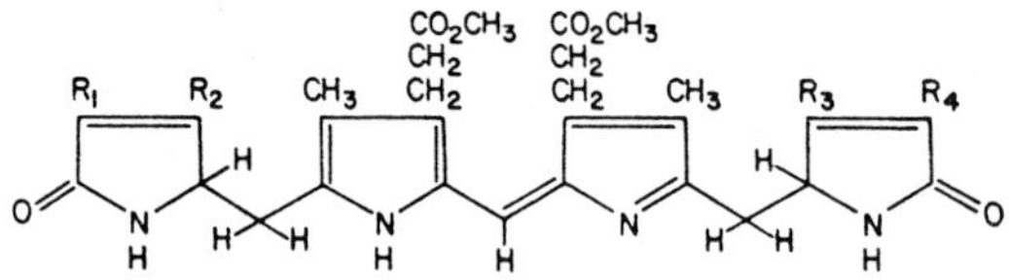

(v)

TABLE IV

End Ring β Substituents and k' Values on μ Porasil in Benzene/Ethanol/Diethylamine (72500/2500/1) of Six Isomeric I-Urobilins (Structure v)

Compound	R_1	R_2	R_3	R_4	Stereochemistry	k'
27	methyl	ethyl	ethyl	methyl	RR'+SS'	2.9
28	methyl	ethyl	methyl	ethyl	RR'+SS'	3.1
29	ethyl	methyl	methyl	ethyl	RR'+SS'	3.5
30	methyl	ethyl	ethyl	methyl	RS'+SR'	4.1
31	methyl	ethyl	methyl	ethyl	RS'+SR'	4.5
32	ethyl	methyl	methyl	ethyl	RS'+SR'	5.1

The inner α carbon atoms of the end rings are asymmetric and so these compounds exist as RR'+SR' forms as indicated in the Table.

The relative retentions of these compounds, shown in

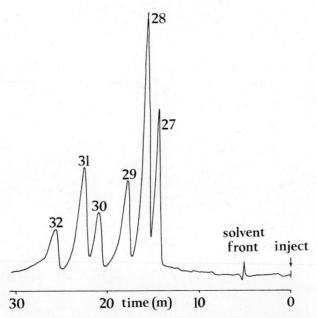

Fig. 2. *Separation of six isomeric i-urobilin dimethyl esters on 30 cm x ⅜ in µ Porasil in benzene/ethanol/diethylamine (72500/2500/1) with detection at 450 nm. Peak numbers refer to compounds in Table IV.*

TABLE V

Relative Retentions of the Six Isomeric I-Urobilins shown in Table IV

Pair of Compounds	Relative Retention
28:27	1.08
29:28	1.14
31:30	1.08
32:31	1.14
30:27	1.44
31:28	1.44
32:29	1.44

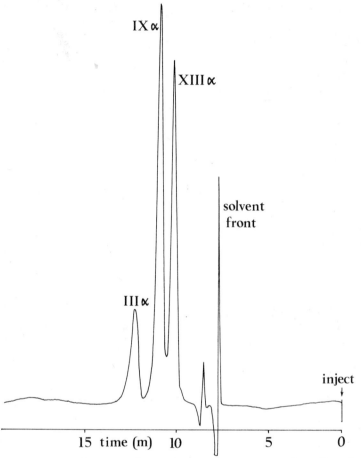

Fig. 3. Separation of three isomeric components of natural stercobilin dimethyl ester on 30 cm x ⅜ in μ-Porasil in benzene/ethanol/diethylamine (1450/50/1) with detection at 450 nm.

Table V, show a striking correlation with their structural relationships. Thus the three pairs of diastereoisomers have identical relative retentions of 1.44 while identical relative retentions of 1.08 and 1.14 are found for XIIIα to IXα and IXα to IIIα compounds respectively in each of the two diastereoisomeric series. No such precise relationships were found in the verdin or violin series.

Examination by HPLC in this system of natural d-urobilin

590 dimethyl ester showed it to consist primarily of SS' i-urobilins IIIα, IXα and XIIIα dimethyl ester with small amounts of the three corresponding RS' compounds. Figure 3 shows the HPLC separation of natural stercobilin dimethyl ester which showed it to consist entirely of the three RR' diastereoisomers, because, although the structure (iv) of these compounds are different from those of the corresponding i-urobilins, no separation could be obtained in this system. The presence of symmetrical isomers in these natural compounds is assumed to be an artefact of their extraction.

(vi)

IIIα $R_1=R_4=$ethyl $R_2=R_3=$methyl
IXα $R_1=R_3=$methyl $R_2=R_4=$ethyl
XIIIα $R_1=R_4=$methyl $R_2=R_3=$ethyl

In each compound class certain generalisations can be made regarding order of retention. With verdins and violins the order was XIIIα> IXα> IIIα but with urobilins it was IIIα> IXα> XIIIα. With violins the nature of the unconjugated end ring had a profound influence i.e. the presence of a vinyl or an ethylidene substituent produced smaller retentions than the presence of an ethyl group and RS' ethylidene compounds had smaller retentions than RR' compounds. With violins also, a vinyl substituent attached to the chromophore produced a smaller retention than an ethyl substituent in the same position. Finally in the urobilin series the RR'+ SS' diastereoisomers had smaller retentions than the RS'+SS' diastereoisomers. It was interesting that in both the urobilin and violin series diastereoisomers were readily separated.

Although quantitative measurements of sensitivity have not yet been made, detection well below the microgram level is easily attained, and since many pigments of clinical importance or compounds readily derived from them can be separated, the means now exists to perform routine analyses of pigments from natural sources.

REFERENCES

1. Stoll, M.S. and Gray, C.H. (1970). *Biochem. J.* <u>117</u>, 271.
2. Petryka, Z.J. and Watson, C.J. (1968). *J. Chromatogr.* <u>37</u>, 76.
3. O'Carra, P. and Colleran, E. (1970). *J. Chromatogr.* <u>50</u>, 458.
4. McDonagh, A.F. and Assisi, F. (1972). *J. Chem. Soc. Chem. Commun.* <u>3</u>, 117.
5. Chapman, D.J., Cole, W.J. and Siegelman, H.W. (1967). *J. Am. chem. Soc.* <u>89</u>, 5976.

HIGH PERFORMANCE LIQUID CHROMATOGRAPHY OF NUCLEOTIDES IN BIOLOGICAL FLUIDS

D. Perrett

*Department of Medicine,
St. Bartholomew's Hospital,
London.*

INTRODUCTION

The recent developments in the field of HPLC have arisen mainly from the interest of gas chromatographers in the possibilities of liquid-solid chromatography. This has resulted in the use of many practices common in gas chromatography such as long columns. Prior to this the amino acid analyser had been investigated and developed for the use of biochemists and clinical chemists. In its present form the automated multiple analysis of protein hydrolysates can be performed in less than thirty minutes. Much practical and theoretical work in this field would also appear applicable to other types of liquid chromatography. In the author's laboratory, amino acid analysis has been performed for over fifteen years and considerable experience has been gained.

When the need for quantitative determination of intracellular nucleotides arose in a study on purine metabolism, HPLC was considered the best approach. The method of Brown (1) for nucleotide analysis using pellicular anion exchangers developed by Horvath et al. (2) gave both the required speed and sensitivity but it involved the use of high pressure equipment. In 1970 Mondino (3) published the results of an experimental study for optimising resin column dimensions for amino acid analysis. He showed that provided the volume of resin required for a given separation was employed the resolution was independent of the dimensions of the column when all other conditions were constant. It was decided to investigate whether the findings of Mondino with regard to cation exchange chromatography could also apply to pellicular anion exchange chromatography of nucleotides. Since the operating pressure at any given flow rate is inversely proportional to the column cross sectional area, the use of wide bore columns should allow the employment of low pressure equipment such as glass columns whilst retaining the high resolution and speed of the pellicular resin methodology. A system employing a column of only 20 cm in length but of the same volume (2.35 ml) as the 3 m x 1 mm column employed by Brown (1) and packed with the same resin is described in this paper.

Apparatus and Methods

A 20 cm x 4 mm glass column (Jobling, Stone, England) was dry packed with AS-pellionex-SAX pellicular anion exchanger (Reeve Angel, Maidstone). All other column fittings were also obtained from Jobling. The column was maintained at constant temperature (usually 75°C) by a circulating water bath. A pulse damped Milton Roy mini pump was employed. The eluent was monitored using a CE212 variable wavelength monitor (Cecil Instruments, Cambridge) fitted with both a 10 μl flow cell and a CE213 Auto Range Unit. The output from the monitor was fed to a 0-10 mv potentiometric recorder.

For the analysis of the mono-, di-, and tri- nucleotides the anion exchange column was eluted with a mixed potassium Phosphate/potassium chloride gradient. The gradient was generated using a simple gradient former, consisting of an open-topped box 20 cm square and of approximately 20 ml capacity constructed with eighth inch perspex sheet. The box was divided diagonally into two separate water-tight compartments by a length of flexible solid polythene tubing. The two chambers were connected through a Technicon micro T-piece, the lower arm of which led to the pump. One compartment was filled with low concentration buffer and the other with the final buffer. The slope of the gradient produced was given by the shape of the flexible tubing.

Nucleotide extracts of cells were prepared using trichloroacetic acid (TCA) as the protein precipitant. For erythrocytes one volume of washed packed red cells with the buffy coat removed were precipitated with 2 volumes of 10% TCA containing sufficient of any ^{14}C-purine base to give approximately 1000 cpm/10 μl of final extracts. The TCA supernatant was then extracted with 2 x 4 vol of water saturated diethyl ether and finally the pH was adjusted to 7 using a few crystals of TRIS. 10 μl of the TCA mixture were also counted in 1 ml of buffer using 2.5 ml of Unisolve I (Koch-Light, Colnbrooke) as the scintillant. The incorporation of this radioactive standard was found to improve overall quantitation by correcting for both recovery and the water volume of the cells.

The above extraction method was slightly modified for use with other blood cells, e.g. platelets, lymphocytes and tissue extracts.

Samples (up to 30 µl) were applied directly to the top of the resin bed via a septum injector using standard microlitre syringes (SGE Ltd., London) without stopping the flow. Nucleotides were identified by their elution times and by cochromatography. It was also found to be possible to stop the pump when a peak of interest was in the flow cell and then make absorbance measurements at different wavelengths. Parallel baseline measurements were also made just before or after the peak in order to make true absorbance readings. Comparison of the wavelength ratios obtained in this manner with tables has proved valuable for identification purposes. Peak areas were measured by the height times half-width method and nucleotide concentrations calculated by relation to standard chromatograms.

RESULTS AND DISCUSSION

As expected the reduction in column length from 3 m to 20 cm whilst still containing the same volume of resin produced a marked decrease in pressure. At a flow rate of 1 ml/min the pressure fell from a reported value of 3000 p.s.i. (4) to less than 50 p.s.i. In agreement with the findings of Mondino (3) the resolution of the anion exchange column was not noticeably different from published results. Although it was not possible with the facilities available to study the effect of varying the column dimensions over a range of constant resin volumes, it has been possible to produce chromatograms directly equivalent to those published by other workers (5). Using a single buffer the adenosine nucleotides have been separated in less than 3 min in a manner identical to that reported by Burtis *et al.* (4) using a 3 m column. A procedure for simultaneous high sensitivity analysis of cyclic adenosine monophosphate and cyclic guanosine monophosphate in the same sample in less than 10 minutes has also been developed (5).

Because of the difficulties of exactly reproducing the

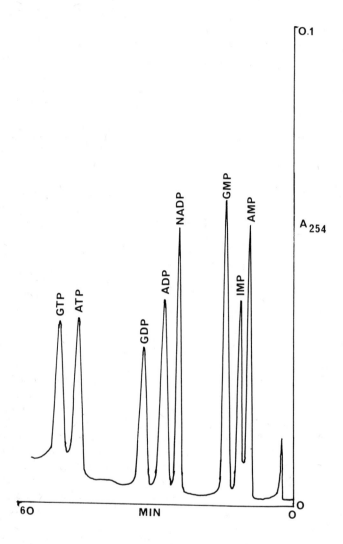

Fig. 1. Separation of standard Nucleotides. Gradient 0.004 M KH_2PO_4 pH 6.5 to 0.18 M KH_2PO_4 + 0.15 M KCl. Flow rate 22 ml/hr. Temperature 75°C. Sample 10 µl containing approximately 3 nmoles of each nucleotide. AMP, ADP, ATP = adenosine 5' mono-, 5'-di, 5'-triphosphate. GMP, GDP, GTP = guanosine 5' mono-, 5'-di, 5'-triphosphate. IMP = Inosine 5' mono phosphate. NADP = Nicotinamide-adenine 5' diphosphate.

gradients employed by other workers, no direct comparison with published gradient chromatograms is possible. Separations of complex nucleotide mixtures both as standard solutions and as biological extracts have been produced which are comparable with the separations published by Brown (1). The number of peaks resolved in complex biological extracts does not appear to be less than the number obtained using much longer columns. Figure 1 shows a typical standard chromatogram containing the nucleotides which are found to be of greatest importance in our studies. A linear gradient from 0.004 M potassium dihydrogen ortho phosphate to 0.18 M potassium dihydrogen ortho phosphate plus 0.13 M Potassium chloride was required to separate the mono-, di- and tri-nucleotides. In order to resolve IMP from both AMP and GMP it was necessary to flatten slightly the initial portion of the gradient and to raise the pH of the first buffer to pH 6.5. The gradient is indicated on Figure 1 by the conductivity line. The column was operated at 75°C and the buffers were pumped at 22 ml/min. Under these conditions the pressure developed was not readable on the pump's pressure gauge, i.e. below 20 p.s.i. In fact for efficient working of the pump and to reduce the problems associated with bubbles in the flow cell, it was found necessary to back pressure the entire system to about 40 p.s.i. by fixing a screw clamp on the p.t.f.e. tubing leading from the flow cell.

Before the technique could be applied to biological materials, suitable extraction techniques must be found. The use of trichloroacetic acid as the protein precipitant was found to lead to fewer losses than perchloric acid. For any material the optimum ratio of sample to precipitant was determined. Because nucleotides are unstable in acid solution, the ATP/ADP ratio, a useful indicator of the quality of the extraction procedure, would rapidly decrease even in acidic extracts stored at -20°C. It was therefore important to neutralise the extracts immediately. Because the direct neutralisation of the TCA extracts with TRIS led to the frontal elution of the nucleotides from the column, it was necessary to remove most of the TCA with ether before the addition of TRIS. Neutralised extracts could then be stored at -20°C for up to two months without substantial degradation. The human erythrocyte provides a readily available biological source of nucleotides. Figure 2 shows a typical chromatogram of human red cell nucleotides extracted with TCA. The use of the CE213 auto range unit can be seen in the ATP peak

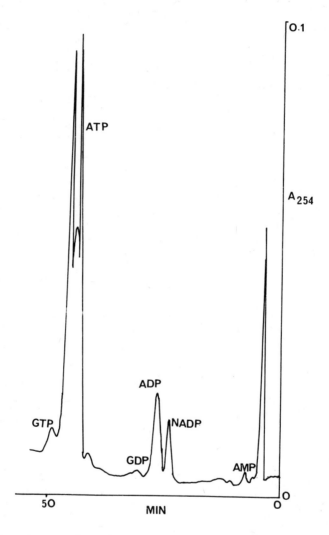

Fig. 2. Separation of nucleotides from TCA extract of normal human erythrocytes. Sample 30 μl of 1:2 extract (see text). Conditions as for Figure 1. Auto range set for × 2 reduction.

where the off scale peak has been reduced by a factor of two. The ATP:ADP ratio being above 8 is indicative of a good preparation for this cell. As can be seen the adenosine nucleotides predominate in the red cell with only trace amounts of the guanosine nucleotides.

For standard solutions the overall precision was similar to those reported by Brown (1) e.g. the coefficient of variation was 2.3%. The accuracy of the quantitation of cellular nucleotides was of course dependent on the peak height. Typically repeated analyses of the same red cell extract gave values of 1395 S.D. ± 70 (n = 10) and 165 S.D. = 10 (n = 10) nmole/ml packed cells for ATP and ADP respectively. Larger errors were obtained for the trace levels of AMP and IMP. Similarly the reproducibility of the extraction technique was determined by repeated sampling and extracting 100 µl aliquots of the same red cell pool and the following values were obtained for ATP and ADP respectively: 1250 S.D. = 75 (n = 19) and 130 S.D. = 15 (n = 16). Figure 3 shows a typical acid extract of human platelets. 20 µl of TCA extract which were equivalent to 0.25×10^6 platelets were injected; sufficient platelets can therefore be obtained from only a small volume of whole blood. The low ATP, ADP ratio in this cell is indicative not of a bad preparation but of the high intracellular concentration of ADP which is important in the aggregation process. HPLC analysis of ADP levels should be of increasing importance in the study of platelet aggregation and already reduced ADP levels in Hermansky-Pulak syndrome have been studied by HPLC (6). The nucleotide profiles and normal values for a wide range of human blood components have been published by Scholar *et al.* (7).

The present equipment although not producing results with the efficiency of the newer micro particular column materials, which can produce the equivalent of Figure 1 in less than 20 minutes, does provide a simple solution to the problems of nucleotide analysis. The present system has now been in routine operation running some thirty to forty chromatograms of biological extracts a week for two years with few operational difficulties. The equipment is readily available and the system is simply constructed and maintained.

Studies on nucleotide metabolism and nucleotide pools are receiving increased attention following the application of HPLC to nucleotide analysis. Metabolic studies are made possible by the non-destructive nature of the analysis which allows the fractionation of the column eluent for radioacti-

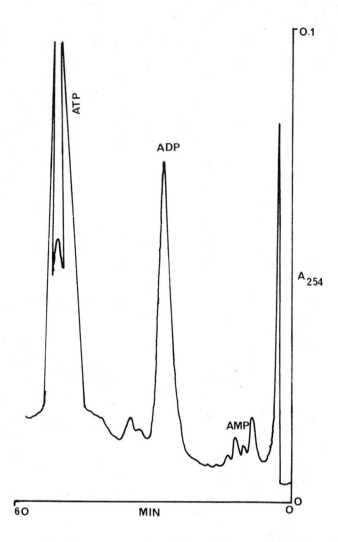

Fig. 3. Separation of nucleotides from TCA extract of normal human platelets. Sample 30 µl = 0.25 x 10^6 cells. Conditions as for Figures 1 and 2.

vity counting. Studies on the effects of drugs on the metabolism of ^{14}C adenosine and adenine have been performed using the present system (8). Nucleotide pools in mouse muscle and the changes in the pools during exercise are currently being studied following the development of suitable extraction techniques. Of particular importance is the recent finding of the first biochemical abnormality associated with an immunological disorder. The finding of a deficiency of the enzyme Adenosine deaminase (E.C. 3.5 4.4), which is important in the control of nucleotide synthesis, in the red cells of children with autosomally inherited severe combined immunodeficiency suggests that studies on nucleotide metabolism using HPLC could play an important part in the understanding of this biochemical abnormality (9).

REFERENCES

1. Brown, P.R. (1970). *J. Chromatog.* 52, 257-272.
2. Horvath, C.G., Preiss, B.A. and Lipsky, S.R. (1967). *Anal. Chem.* 39, 1422-1428.
3. Mondino, A. (1970). *J. Chromatog.* 50, 260-273.
4. Burtis, A.C., Munk, M.N. and MacDonald, F.R. (1970). *Clin. Chem.* 16, 667-676.
5. Perrett, D. (1976). Submitted for publication.
6. Rao, G.H.R., White, J.G., Jachimowicz, A.A. and Witkop, C.J. (1974). *J. Lab. Clin. Med.* 84, 839-850.
7. Scholar, E.M., Brown, P.R., Parks, R.E. and Calabresi, P. (1973). *Blood,* 41, 927-936.
8. Dean, B.M. and Perrett, D. (1975). Submitted for publication.
9. Giblett, E.R., Anderson, J.E., Cohen, F., Pollara, B. and Mevwissen, H.J. (1972). *Lancet,* 2, 1067-1069.

HIGH PRESSURE CHROMATOGRAPHIC SEPARATION OF SOME BIOGENIC AMINES AND DERIVATIVES IMPLICATED IN THE AETIOLOGY OF SCHIZOPHRENIA

T. J. X. Mee and J. A. Smith

Postgraduate School of Studies in Pharmaceutical Chemistry, University of Bradford.

Abnormal methylation has long been associated with schizophrenia (1) leading to an accumulation of data involving both methylated indolealkylamine derivatives and catecholamines. Thus, the indole melatonin (NAMT) (N-acetyl-5-methoxytryptamine) has been implicated in the disease (2,3,4,5). NAMT is formed uniquely in the pineal gland from N-acetyl-5-hydroxytryptamine (NAHT), the methylation being catalysed by hydroxyindole-O-methyl transferase (HIOMT). Part of the NAMT is then hydroxylated in the liver to 6-hydroxymelatonin (6HNAMT).

The levels of NAMT in human blood and urine are so small that current commercial detectors for high pressure liquid chromatography are not sufficiently sensitive. Thus, a radioimmune assay (RIA) is used which is sensitive to 5 pg/ml of blood. However, other molecules related to NAMT could cross react with the antibodies. This paper describes the separation of NAMT from possible cross reacting indolealkylamine derivatives and also the use of HPLC in conjunction with RIA to identify the immune reactive material.

Furthermore, pineal HIOMT has been shown to catalyse the formation *in vitro* of N-acetyl-β-3,4,-dimethoxyphenethylamine (NADMPE) by an abnormal methylation of either N-acetyl-β-3-hydroxy-4-methoxyphenethylamine or N-acetyl-β-4-hydroxy-3-methoxyphenethylamine (6). Other abnormally methylated catecholamines associated with the disease are β-3-hydroxy-4-methoxyphenethylamine (3HMPE) and β-3,4-dimethoxyphenethylamine (DMPE) together with their acidic metabolites 3-hydroxy-4-methoxyphenyl acetic acid (3HMPAA) and 3,4,dimethoxyphenyl acetic acid (DMPAA).

Some preliminary studies on the high pressure liquid chromatographic separation of these and related catecholamines and their derivatives are reported in view of the recent revised interest in DMPE (7).

Methods and Materials

The materials were supplied by Sigma Chemical Company. The separations were effected at 25°C on a Waters ALC/200 Liquid Chromatograph using a UV detector (254 nm) and a solvent programmer attachment (Model 660).

Separation of the Indoles and their Derivatives

Reverse-phase liquid chromatography with a C_{18} stationary phase (Corasil Bondapak) was employed Column length was usually 6 ft with water jacket at 25°C. The mobile phase was a methanol/0.2 M phosphate buffer pH.6.0 mixture starting at 5% methanol and stepping up to 19% after 7 minutes and up to 43% after 19 minutes. The flow rate was 2 ml/min and chart speed 0.5 cm/min (x16 attenuation).

Injections were 20 µl of a solution in water or human serum (protein precipitated) of a mixture of indoles containing approximately 0.5-1.0 mg/ml of each.

Separation of 3HMPAA and DMPAA

Reverse-phase liquid chromatography was again used with a 4 ft Corasil C_{18} Bondapak column (water jacket at 25°C). The mobile phase was 3% methanol in 0.25% acetic acid at a flow rate of 2 ml/min. Injection was 10 µl of aqueous solution containing 2 mg/ml of each. Chart speed was 1 cm/min (x08 attenuation).

Separation of NADMPE and NAMT

Adsorption chromatography was employed with a 4 ft Corasil II column (water jacket at 25°C). The mobile phase was chloroform at 1 ml/min. Injections of 10 µl of an alcoholic solution containing 2 mg/ml of each. Chart speed was 1 cm/min (x16).

Separation of Methylated Catecholamines

A C_{18} Corasil Bondapak column (8 ft with water jacket

at 25°C) was used. The mobile phase was 1% ammonium carbonate solution with either 15% or 25% acetonitrile at flow rates of 1-2 ml/min. Chart speed was 1 cm/min (x16). Injection of 25 μl of an aqueous solution of 0.5 mg/ml of each.

TABLE I

Relative Elution Volumes of Melatonin and Related Compounds as determined by HPLC

Compound	C_{18} Bondapak 6 ft		Corasil II 4 ft
	5% Methanol in 0.2 M Phosphate Buffer (pH 6.0)	15% Methanol in 0.2 M Phosphate Buffer (pH 6.0)	Chloroform
6HNAMT	>2.0 (Very Broad Peak)	1.0+	
NAMT	Not eluted	1.4	1.0†
NADMPE	Not eluted	1.4	0.5
NAHT	1.0*	0.7	
5MTP	0.7	0.2	
5HTP	0.4	0.2	
5MT	Not eluted	>2.0	
5HT	1.0	0.7	

*Retention Volume at 2 ml/min 14 ml.
+Retention Volume at 2 ml/min 6.5 ml.
†Retention Volume at 1 ml/min 12.5 ml.

RESULTS AND DISCUSSION

Table I indicates that six possible cross reacting indoles may be separated on a C_{18} column from NAMT in human serum. This is effected by stepped elution with methanol/ 0.2 M phosphate buffer pH 6.0 solutions commencing with 5% methanol for seven minutes, 19% methanol for a further 12 minutes and 43% for another eight minutes. The indoles elute in order; 5-hydroxytryptophan (5HTP), 5-methoxytryptophan (5MTP), 5-hydroxytryptamine (5HT), N-acetyl-5-hydroxytryptamine (NAHT), 6HNAMT, NAMT and 5-methoxytryptamine (5MT).

This separation is of great interest to those using RIA

to measure NAMT and related indoles in biological fluids. Although we use an assay (8) which is quite specific for NAMT, the metabolite most likely to cross react with the antibody is 5MT. However, the antibody in the RIA method used by Grota and Brown (9) to measure NAHT cross reacted seriously with NAMT. Furthermore, 5MT competes to a great extent in the RIA of 5HT (9,10).

Thus, the separation of these indoles and no doubt other related molecules by HPLC provides a valuable method for eliminating cross reacting metabolites in preparation for RIA. The protein precipitated serum or extracted serum may be passed through the column and 0.5 ml samples collected. After drying, the samples may be assayed by RIA.

In addition, it provides evidence to support the identity of the immuno reactive material. Tritiated NAMT can be added to the mixture of indoles and injected on to the column under the same conditions. The collection of 0.5 ml samples will show that the plot of counts per minute against elution volume coincides with the chart record of the cold NAMT as measured by the UV detector.

Also, the aliquots due to NAMT will react with the antibody whereas the other aliquots will not. Finally the presence of only one peak when tritiated NAMT is injected on to the column provides evidence for stability of the labelled NAMT.

A similar combination of HPLC and RIA was recently reported (11) in separating cross-reactants from DMPE and in establishing the identity of the immuno-reactive material.

Table I also shows that NADMPE has the same elution volume as NAMT in the methanol/buffer system. However, using adsorption chromatography (Corasil II) and chloroform, the two may be separated. Since NADMPE produces behavioural changes in animals and is related to DMPE, the former is of interest in psychiatric biochemistry. However, sensitivity (500 ng) on the column prevents the direct analysis of this material in human serum.

In the preliminary studies on the separation of the β-phenethylamine derivatives, three groups were investigated. Two of the groups, the methylated dopamines and the methylated epinine derivatives, were chromatographed on a C_{18} Bondapak column (8 ft) using 15% or 25% acetonitrile in 1% ammonium carbonate solution as solvent.

Table II indicates that in both groups the monomethylated derivatives (3MT/4MPE and N3MPE/N4MPE) were not resolved.

TABLE II

Relative Elution Volumes of Methylated Catecholamines and Derivatives as determined by HPLC

Compound	C_{18} Bondapak 8 ft		C_{18} Bondapak 4 ft
	15% Acetonitrile in 1% Ammonium Carbonate	25% Acetonitrile in 1% Ammonium Carbonate	3% Methanol in 0.25% Acetic Acid
DMPE	1.5	1.2	–
3MT	1.0*	1.0+	–
4MPE	1.0	1.0	–
NDMPE	Not eluted	1.7	–
N3MPE	1.0†	1.0ϕ	–
N4MPE	1.2	1.0	–
DMPAA	–	–	1Δ
3HMPAA	–	–	0.3

*Retention Volume at 1 ml/min 6.7 ml
+Retention Volume at 1 ml/min 6.1 ml
†Retention Volume at 1 ml/min 11.0 ml
ϕRetention Volume at 1 ml/min 7.5 ml
ΔRetention Volume at 2 ml/min 19.0 ml

However, 3MT and 4MPE could be partially separated after three recycles using 5% acetonitrile in 1% ammonium carbonate solution at 1 ml/min (25°C).

On the other hand, Table II shows that 3MT or 4MPE could be separated from DMPE and that N3MPE or N4MPE may be resolved from NDMPE. These results are similar to those reported by Riceberg and Van Vunakis (11), who used a diphenyl Corasil column and methanol in 0.1% ammonium carbonate solution (65:35) as solvent. They were able to separate 3MT from DMPE but 3MT and tyramine remained unresolved.

Also, we found the compounds in either of these two groups could not be resolved using ion exchange (AX/Corasil 2 ft) with 0.2 M borate buffer pH 9.0 or with 0.2 M phosphate buffer at pH 7.0. Nor was separation achieved on a Vydac TM column (4 ft) using chloroform or chloroform/methanol mixtures.

Finally it is seen from Table II that 3HMPAA may be separated from DMPAA using 3% methanol in 0.25% acetic acid and a C_{18} column (4 ft).

The sensitivity of the method for these compounds is around 100-500 ng on the column. Combined with RIA, the sensitivity can be increased 1000 fold (11). For direct analysis of serum, a direct HPLC method must depend on a more sensitive method such as spectrofluorimetric or polarographic detection.

REFERENCES

1. Kety, S.S. (1965). *Int. J. Psychiat.* 1, 409-415.
2. McIssac, W.M., Khairallah, P.A. and Page I.M. (1961). *Science,* 134, 674-675.
3. Jones, R.L., McGeer, P.L. and Greiner, A.C. (1969). *Clin. Chem. Acta.* 26, 281-283.
4. Greiner, A.C. (1970). *Can. Psychiat. Assoc.* 15, 433-447.
5. Hartley, R., Padwick, D. and Smith, J.A. (1972). *J. Pharm. Pharmac.* 24, 100P-103P.
6. Hartley, R. and Smith, J.A. (1973). *Biochem. Pharmac.* 22, 2425-2428.
7. Braun, G., Kalbhen, D.A., Muller, J. and Vahr-Martiar, H. (1974). *Arch. Psychiat. Nervenkr.* 218, 195-210.
8. Arendt, J., Paunier, L. and Sizonenko, P. (1975). *J. Clin. Endocrinol. and Metab.* 40, 347-350.
9. Grota, L.J. and Brown, G.N. (1974). *Can. J. Biochem.* 52, 196-202.
10. Ranadive, N.S. and Sehon, A.H. (1967). *Can. J. Biochem.* 45, 1701-1710.
11. Riceberg, L.J. and Van Vienakis, H. (1975). *Biochem. Pharmac.* 24, 259-265.

SEPARATION OF CATECHOLAMINES AND THEIR
METABOLITES BY HIGH SPEED LIQUID CHROMATOGRAPHY

J. Jurand

*Wolfson Liquid Chromatography Unit,
Department of Chemistry, University of Edinburgh.*

INTRODUCTION

A novel reverse-phase chromatographic technique has been employed in which a bonded packing material is used with a surface active agent added to the eluent. The technique is superior to liquid-solid, ion-pair partition or ion-exchange chromatography in terms of column efficiency, speed, sensitivity and its ability to resolve compounds of interest including catecholamines, their 3-O-methyl derivatives, L-Dopa and related compounds is described.

Two modifications were made to the conventional reverse-phase chromatography method which employs octadecyl silica as stationary phase in order to improve the chromatography of the catecholamines and related compounds. First a conventional octadecylsilyl (ODS) silica (manufactured in the Wolfson Liquid Chromatography Unit; WLCU) was treated to substitute remaining silanol groups with $-Si\ Me_3$ (1) and second a detergent was added to the mobile phase. The addition of small concentrations of detergent was found to be essential for adequate retention of relevant compounds in the reverse-phase system.

Experimental

The high speed liquid chromatograph was constructed in the laboratory. It consisted of a high pressure gas driven pump (Haskel, Burbank, California, U.S.A.) and a variable wavelength UV detector (Cecil Instruments, Cambridge).

Columns were 5 mm wide and 125 mm long of internally polished stainless steel. They were terminated by 0.8 μm silver membranes (Flotronics Ltd., Pennsylvania, U.S.A.) sandwiched between two 12 μm porosity metal frits (B.S.A. Ltd., Birmingham). Column fittings were made according to a design described elsewhere (2). Columns were operated at room temperature.

The packing material developed by A. Pryde (WLCU) was prepared by reacting 6-7 μm spherical silica (WLCU) with octadecyl silane. Any residual silanol groups were subsequently blocked with trimethylsilyl groups (1) to form an ODS/TMS chemically bonded silica.

Columns were slurry packed at 3000 psi using methyl iodide as a balanced density solvent.

Mobile phases consisted of methanol-water, acetonitrile-water, propanol-water or dioxane-water mixtures acidified with 0.04%-0.2% (by vol) of sulphuric acid and incorporating one of three surface active components: sodium lauryl sulphate (SLS), sodium dodecane sulphonic acid (SDS) and sodium dodecyl benzene sulphonic acid (SDBS).

The following pure reference compounds were obtained from Sigma Ltd. or from Hoffmann La Roche: noradrenaline (NA), adrenaline (A), L-Dopa (LD), dopamine (DA), normetadrenaline (NMA), metadrenaline (MA), 3-methoxytyramine (MDA), tyramine (TYR), D,L-alpha-methyl Dopa (α-MD), homovanillic acid (HMV), 4-hydroxy 3-methoxymandelic acid (HMMA or VMA) and 3-4-dihydroxyphenylglycol (DPG).

Results

In the reverse-phase chromatography the catecholamines are eluted before their 3-0-methyl derivatives. This is the reverse order to that obtained in liquid-solid and in ion pair chromatography.

Quantitative elution can be achieved only with the use of acidic eluents but under these conditions most of the components of interest are either unretained or slightly retained and are then eluted as broad peaks.

Addition of a small quantity of detergent to the mobile phase has a dramatic effect on retention and peak symmetry and has been found essential in order to achieve satisfactory retentions and good column efficiency.

The study on the effects produced by the three different detergents using 11/29 v/v methanol-water mixture (pH 2.3) containing 0.04% (by vol) of sulphuric acid has shown that individual detergents have different effects on the retention times of individual components. The highest selec-

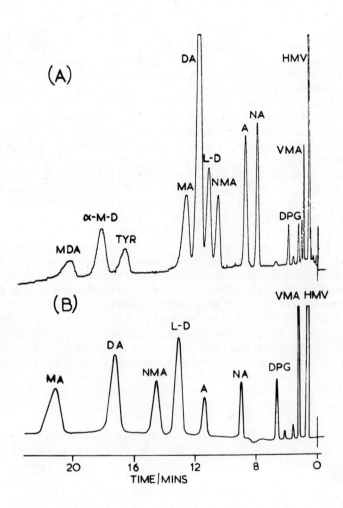

Fig. 1. HSL-chromatogram of catecholamines and related compounds. Column 125 x 5 mm; packing ODS/TMS silica (WLCU); detector, UV photometer at 280 nm, sensitivity 0.02 AUFS; eluent (A) methanol-water-sulphuric acid SDS 27.5/72.5/0.04/ 0.02, (B) acetonitrile-water-sulphuric acid-SLS 11/89/0.04/ 0.01. Solute identification see text. Note reversal of order of elution of L-Dopa and NMA. Injected 10-100 ng of each component (40 ng (A) and 25 ng (B) for NA).

tivity of separation was obtained with SLS, concentrations as low as 0.005% (by weight) being effective. Optimal separation with SDS was obtained at a concentration of 0.02% and with SDBS at a concentration of 0.04%. At higher detergent concentrations column efficiency deteriorated and retentions of the 3-methoxytyramine, tyramine and of alpha methyl dopamine became excessive.

The second parameter which determines the selectivity and the efficiency of separation is the nature of the organic hydrophilic solvent. Methanol and acetonitrile gave the best column efficiencies, whereas propanol and dioxane gave poorer performance.

Figure 1(A) illustrates the separation obtained with methanol-water containing SDS and Figure 1(B) illustrates the separation obtained with acetonitrile-water containing SLS. The better selectivity obtained on Figure 1(B) results mostly from the change of the hydrophilic solvent. Both detergents were used at their optimal concentrations. The order of elution using methanol and acetonitrile is different since L-Dopa is eluted before normetradrenaline with acetonitrile and after it with methanol. Although elution of all components is accelerated with increasing concentrations of acetonitrile in the mobile phase, L-Dopa is more drastically affected by this increase than noradrenaline, adrenaline or normetadrenaline and therefore its position on the chromatogram in relation to other components is critically affected by very small changes in acetonitrile concentration.

The individual retention time is also dependent upon pH. During the study concentrations of sulphuric acid were used between 0.04 and 0.2% (0.004 to 0.02 M H_2SO_4). At the higher concentrations of acid the retention times were generally shortened but those of L-Dopa and D,L-α-methyl-Dopa were little affected.

Urine Analyses

In a preliminary study of urine analysis, direct injections of 5-25 μl untreated urine were made. Urine specimens were collected from normal individuals, and from a Parkinsonian patient treated either with L-Dopa (Hoffmann La Roche) and with Madopar (L-Dopa plus benserazide, Hoffmann La Roche).

Figure 2 shows the differences obtained between the normal urine (C) and pathological urines obtained from the patient when in L-Dopa treatment (A) and Madopar treatment (B),

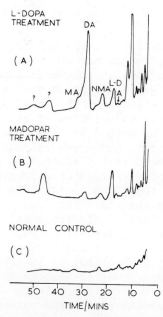

Fig. 2. HSL-chromatogram of urines from Parkinsonian patient treated with L-Dopa (A) and Madopar (B), and of normal urine (C). Sample 25 µl (A) 5 µl (B) and (C) untreated urine. Chromatographic conditions as for Figure 1 except eluent was acetonitrile-water-sulphuric acid-SDS 15/85/0.06/0.01.

the solvent system being acetonitrile-water. The pathological urines show many more and substantially larger peaks. In general they occur at retention times corresponding to those of the catecholamines, their methyl derivatives, L-Dopa and other related compounds.

In addition, the urine obtained after L-Dopa treatment contains more of the component with a retention time corresponding to dopamine, and less of the component corresponding to L-Dopa when compared with the urine obtained after Madopar treatment. This is confirmed using eluents of several different compositions. At this stage positive identification of these peaks cannot be assured and although running of the analyses in different solvent mixtures can help in identification certain identification must be carried out by an independent method such as mass spectrometry.

DISCUSSION

The new chromatographic technique described above for the separation of catecholamines and their metabolites compared favourably with the ion-pair method previously described by Persson and Karger (3). The use of a chemically bonded stationary phase eliminated the need to use precolumns to saturate the eluent, and eliminates bleeding from the column. In addition, the method is more flexible, since the composition of the mobile phase can very easily be altered in respect of the concentration of organic solvent, the pH and the nature and concentration of detergent, all of which significantly affect the chromatographic performance.

It is also clear that the technique has considerable potential for the quantitation of components of interest in urine. Adrenaline and noradrenaline for example can be detected at the level of a few nanograms per injection. The preliminary results indicate that it is certainly worthwhile extending this study in order to elaborate a quantitative technique which could monitor the effects of L-Dopa and similar drugs on the metabolic profile of catecholamines.

REFERENCES

1. Knox, J.H. and Pryde, A. (1975). *J. Chromatogr.* 112, 171-188.
2. Knox, J.H. (1975). *Chem. Ind.* (London), 29-34.
3. Persson, B.A. and Karger, B.L. (1974). *J. Chromatogr.* 12, 521-528.

PROBLEMS IN SEPARATING URINARY METADRENALINES

J. P. Leppard, A. D. R. Harrison and E. Reid
*Wolfson Bioanalytical Centre,
University of Surrey.*

There is need for reliable and simple ways to estimate normetadrenaline (normetanephrine, NM) and metadrenaline (metanephrine, M). Published methods (1) are all intricate, and have commonly been reported without comparison with a previous method. In general the urine sample, hydrolyzed to liberate NM and M from conjugates, has to be processed to remove noradrenaline, adrenaline and other interfering materials; finally the NM and M are either separated by chromatography or are distinguished during off-line measurement, which is commonly fluorimetric.

Although the urinary content of NM and M is low (~0.1 and ~0.05 mg l^{-1} respectively) and their UV absorbance (maximal at ~279 nm) is poor, in principle it was a reasonable aim to use HPLC with a UV detector to estimate NM and M. It seemed obligatory that the urine be concentrated, perhaps one hundredfold, and cleaned up initially.

Sample Preparation and Reference Method

Various approaches to working up the urine samples (1, 2,3) were considered, taking account of the instability of NM and M (both of which are basic) at alkaline pH and that at physiological levels irreversible losses could be serious. Solvent extraction and chromatography on a weak cation-exchanger were each rejected because of unreliable recoveries. The procedure adopted (based upon that of Kahane and Vestergaard, 4,5) was to prepare a cationic concentrate containing NM and M with a strong cation-exchanger, after acid hydrolysis of the urine to split conjugates, as shown in the following scheme:

TABLE I *Exploratory HPLC Work with an Authentic Mixture of*

	1. Micropak-SI 10 2. SLC-3 3. SLC-11 (Adsorptive)	Micropak-CN (Bonded Phase)	Micropak-CH (Reverse Phase)
Solvents/buffers tried *and NM and M behaviour*	Up to 10% conc. NH_3 in, e.g., methanol, or butanol or isopropanol – *no separation*	Methanol or isopropanol + pH 8.5 Na acetate – *NM gave poor peak*	Isopropanol + water, suitably 1:1 by vol. – *separation*
Attempts to optimize use of one of the above solvents/buffers or of a variant			
Problems encountered/variations tried in connection with urinary concentrates (gradient elution)			Awaits trial with urine but
Remarks	Abandoned	Abandoned	–CH phase prone to strip off

NM and M, Usually Without a Gradient, on 25 or 50 cm Columns

Aminex-7 (Cation-Exchange)	Pellionex (Pellicular Cation-Exchange)	Partisil (Microparticulate Cation-Exchange)	Spherisorb-ODS (Reverse Phase on Spherical Beads)
Ammonium phosphate buffers, up to 0.5 molar, pH 5.5 or 7.5 – *no elution*	Ammonium phosphate or Na formate or citrate buffers, pH 2.5 upwards – *no separation*	Na citrate or (keeps better) formate buffer, pH 2.5-8 – *separation, with some tailing if no gradient or if pH>7; small admixture of methanol→ shorter R_T*	Aqueous methanol or (better) acetonitrile, + ammonium carbonate to 0.16% – *separation*
	Grad., pH ∿9.5 gly – *?hopeful* Borate, pH 7 or 10 – *no elution*	pH other than ∿5.5 (longer R_T?) – *no help* Borate, or NH_4^+ – *poorer separation* Urea – *?helpful*	Omit ammonium carbonate – *poor peaks* Urea – *may be pptd. by acetonitrile*
		UV-absorbing "junk" – *diminish by initial elution with, say, 10% methanol.* "M" sometimes "vanishes" – *mercaptoethanol ?helps*	UV-absorbing "junk" – *diminish by initial elution with 10% or (better) 20% methanol* "M" sometimes "vanishes"
Abandoned	Abandoned R_T↓if pH>7 HC-Pellionex worse than HS – *no elution*	Intractable contamination of the NM and M, least at pH ∿5.5. Ion-exch. groups prone to strip off, even if pH never alkaline.	Intractable contamination of the NM and M, even if isocratic. ODS prone to strip off, even if pH little above 8

Frozen and thawed urine, *ca.* 25 ml fraction:
Filter, dilute to 100 ml, adjust to pH 1 with HCl.

    Hydrolyse | $100°$, 30 min.

    Adjust to pH 6 with NH_4OH, add 1 g urea, dilute to 200 ml, load onto 6.5 Dowex 50 column (NH_4).

    Wash column with water and (optional, see text) 50% v/v methanol. Elute with 10 ml 75% methanol containing 1.25 M NH_3. Evaporate to dryness.

    EITHER   OR

Chromatograph on cellulose phosphate. Separated NM and M peaks are then assayed as lutidines fluorimetrically.   Redissolve in small volume of methanol for HPLC

Recoveries, typically 60% for both NM and M, were routinely determined by counting samples spiked with a tracer amount of DL, (7-^{3}H) NM and/or M (New England Nuclear) before and after the sample preparation steps. During the drying down of the ammonia from the eluate, in our experience noradrenaline and adrenaline (but not NM or M) are destroyed (2,4). The aqueous methanol wash of the column and especially the initial removal of solids that fail to redissolve when frozen urine is thawed help free the sample from some material which can jeopardize HPLC and fluorimetry. Unlike other procedures which we have attempted to set up as reference methods, this multi-stage procedure has given fair reproducibility, if meticulously followed, and reasonable agreement with the rather wide range of normal values published for NM and M, our values being typically 150 µg/24 hr NM and 100 µg/24 hr M. For the fluorimetry we prefer the ferricyanide-ascorbate method of Smith and Weil-Malherbe (6) to later variants. Amines such as tyramine do not react.

Approaches Tried for HPLC Assay

    Methanolic solutions of authentic NM and M (Sigma) 2 µg, were used in exploratory experiments with silica-based packings (Table I). Of the totally porous packings, only with Micropak-CH was separation of NM and M achieved, with aqueous isopropanol; NM emerged before M, as in the other separations outlined below. The Micropak-CH column subsequently lost its resolving power - a persistent phenomenon

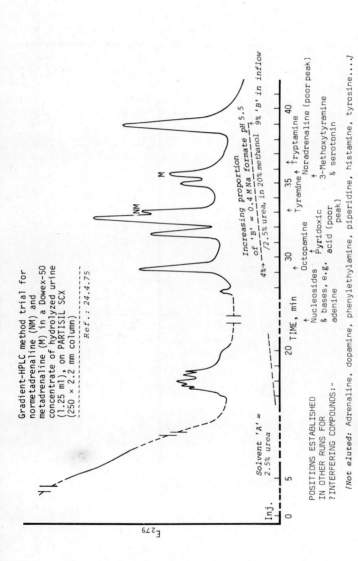

Fig. 1. Separation of NM and M on a Partisil-10 SCX (10 μm) column (Reeve Angel/Whatman) with a 5 cm pre-column containing a pellicular cation-exchanger, Pellionex HS (same supplier), in a Varian 4200 HPLC apparatus with a UV detector (Cecil Instruments) under constant-flow conditions (1 mL/min; ~850 psi). The sample (5 μL) of urinary concentrate was introduced by stopped-flow (septum-less) injection.

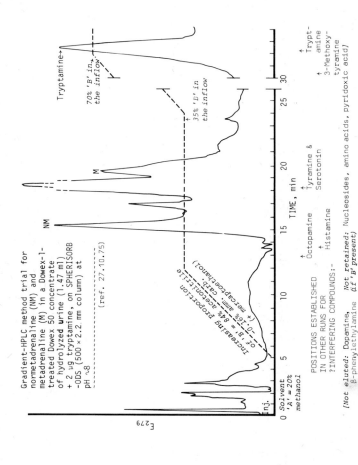

Fig. 2. Separation of NM and M on a uniform-bead (10 μm) reverse-phase silica-based packing (PhaseSep) under constant-flow conditions (1 ml/min). The 5 cm pre-column contained the same packing.

throughout the present study.

Cation-exchange separation was next attempted (Table I), notwithstanding the similarity between the amino pK of NM, namely 9.56, and that of M, 9.71[7]. Aminex-7 (polystyrene-based) was too retentive. A pellicular packing failed to separate NM and M although a further study of conditions might have been rewarding (P. T. Kissinger, personal communication).

A micro-particulate packing (Partisil-10 SCX, 10 μm) in the form of pre-packed columns gave encouraging results. Isocratic pilot runs showed tolerably good individual peaks for NM and M, with some tailing, at pH 2.5 or 5.5 (shorter R_T), and also at pH 8.0 where there was a puzzling lengthening of R_T with a broadened M peak. Most further work was done at pH near 5.5, with formate rather than citrate which tends to grow moulds. The latter hazard to expensive columns can be minimized by having 0.05% sodium azide present in the buffer.

When hydrolyzed urine samples concentrated by the Dowex-50 procedure were tried, usually with urea added, it became apparent that the NM-M region was severely contaminated by adventitious UV-absorbing material. This could be largely removed by initial elution with water (Fig. 1). Urea was added in the hope of minimizing adsorptive effects (4). Gradient elution with sodium formate buffer then gave a somewhat complex pattern of peaks which, although rather variable, always showed UV-absorbing material in the NM position and usually showed some in the M position. These positions were confirmed in several experiments by spiking or, more conclusively, by adding tracer amounts of tritium-labelled NM and M before loading and counting the ^{3}H in the collected fractions over and around each peak. The ^{3}H recovery was typically 80%. Any slight mis-matching between the presumptive NM and M peaks from urine and the isotopic label was attributable to decomposition of the tritiated NM and M stock solutions even at -20° (probably circumvented in recent work by storing in liquid nitrogen). At least the UV pattern of Figure 1 showed evidence of NM, as an unseparated peak in the descending portion of the ?tyramine peak. Octopamine can serve as internal standard.

In later experiments 10% methanol was used as solvent "A", which appeared to improve the initial elution of interfering UV-absorbing material. When the gradient was commenced at 15 min (4,5) no NM secondary peak was seen and a single peak was obtained.

Although quite sharp "NM" and "M" peaks were obtainable with Partisil SCX there was an excess of UV-absorbing material relative to the amount of NM and M found by the independent fluorimetric assays (reference method). The "excess UV" values - at least twice the expected values - increased to about 10 times in later experiments where the gradient was steepened. The excess material in either peak was not fluorigenic in the reference assay as applied to collected fractions from some early HPLC runs. Inferences about identity can be drawn from HPLC positions established for certain authentic compounds as shown below the abscissa in Figure 1. Tyramine evidently may be present in the "NM" peak, although runs with authentic compounds have shown some tendency for its position to vary relative to NM. With urine it may swamp NM, its daily output being high (> 1 mg).

For the "M" peak, the contamination could be by serotonin or, more probably, 3-methoxytyramine (daily output ~0.5 mg) - which does not react in the reference assay method, whereas serotonin reacts slightly but would have been destroyed by the pH 1 hydrolysis (6). Although the "NM" peak was usually large, occasionally there was an inexplicable absence of a peak in the M region. This apparent loss might be avoided by the use of mercaptoethanol (1), but it is unpleasant to handle, even at 0.05% concentrations.

Other suspected contaminating compounds ran differently from authentic NM and M, as is indicated at the foot of Figure 1; some even resisted elution. The range of cationic compounds tested included amphoteric compounds; but it was unlikely that these were interfering since high UV absorbances were still obtained for the "NM" and "M" peaks if, before or after Dowex-50, the sample were put through a Dowex-1 (anion-exchange) column, or if the Dowex-50 column were washed with sodium acetate solution - a procedure that is reported to remove amino acids (4).

No success has been achieved in trials of various stratagems to reduce the UV absorbances of the "NM" and "M" peaks to plausible levels. Trial of pH 3 or pH 8 in place of pH 5.5 gave confusing UV patterns. Pre-washing of the urine with heptane did not help, nor did omission of the acid hydrolysis.

A further serious problem with Partisil has been deterioration of the column, even after only a dozen runs. The deterioration was manifested by shortened retention times and broad peaks for NM and M, if still separable. It occurred even with a lightly used column (cost £115) that (as recommended by the manufacturers) had never been exposed to a pH above 7.6 or to strong electrolyte solutions.

With Spherisorb-ODS reverse phase packing (Table I), NM and M were separable using aqueous acetonitrile at a pH just above 8; an initial flush with 20% methanol removed some adventitious UV-absorbing material from urinary concentrates. Figure 2 shows a chromatograph of a urine extract. Although in this run the Dowex-50 concentrate was passed through Dowex-1 to remove interfering amphoteric compounds, in other runs the amount of UV-absorbing material in the NM and M regions was lower than that illustrated, yet this material was grossly excessive (even more than 10-fold) in relation to the true urinary contents of NM and M as established by the reference method. Various authentic compounds as in the above-mentioned Partisil experiments were investigated as possible contaminants (Fig. 2); evidently only octopamine (p-hydroxyphenylethanolamine) might conceivably contribute to "NM". Tyramine can serve as an internal standard.

Column performance quickly deteriorated. With a fresh column, loss of performance was observed after relatively little use with only authentic compounds. Other fresh columns (with a different batch of Spherisorb, 14/185 in place of 14/137/2) showed virtually no reverse-phase properties even at the outset. The reason for this deterioration might be the need to work at a pH slightly above 8, albeit still about 1.5 pH units below the pK values for NM and M.

DISCUSSION

The deterioration observed with each of three types of bonded-phase packings, from different manufacturers, was a disturbing finding, particularly surprising in the case of Spherisorb-ODS where the ODS chain is bonded to the silica core via C-Si rather than C-O-Si bonds. The deterioration was not due merely to running urinary concentrates, and in any case a pre-column was routinely included (see Legends to Figures 1 and 2) as a precaution which, in common with Dell (8), we consider to be highly desirable for biological fluids. Storage of columns in unsuitable media cannot account for the trouble, as media containing an organic solvent were deliberately used for storage (except in the case of Partisil-SCX) and alkaline pH values were avoided. It was, however, impossible to avoid slight alkalinity in the Spherisorb runs, and the need for this pH may contra-indicate the use of currently available ODS or similar bonded-phase columns for organic bases.

Although NM and M were separated with the SCX or ODS packings, urinary concentrates contained adventitious UV-absorbing contaminants eluting with NM or M. The "NM" or "M" material obtained with the ODS packing might fractionate differently if re-run on the cation-exchanger, giving purer NM and M peaks, although this would not be suitable for routine assay of urinary NM and M.

The content of NM and M in urine is such that, even with, say a 100-fold concentration step such as the Dowex-50 procedure, the HPLC detector would have been operating near its sensitivity limit. The original concept of an HPLC method remains valid, but so far the contamination problem has proved intractable, even for phaeochromocytoma urine. Although independent assays in our laboratory by a GC method (F. Bubb, W. Wegg and M. Sandler, personal communication) have not given values consistently in accord with those by our fluorimetric reference method, they have confirmed that the disagreement with HPLC assay is indeed due to a fault in the latter rather than in the reference method.

We conclude, then, that serious difficulties beset the development of an HPLC end-step for simplifying the assay of urinary metadrenalines. The same view has been reached by P. T. Kissinger (personal communication) using polarographic detectors. It is known that NM and M can be separated by ion-pair HPLC (9), but this approach seems less likely to lead to a satisfactory assay method for urine than a reverse-phase approach with addition of detergent (J. Jurand, this Symposium).

The present work has also delineated the behaviour of various biogenic amines and metabolites in cation-exchange and reverse phase HPLC systems, and has disclosed endogenous peaks in urinary concentrates which are tolerably reproducible and can, in some instances, be tentatively identified. For example, the cation-exchange system shows an endogenous peak which is close to the position of octopamine (Fig. 1).

Whatever method for NM and M might eventually emerge from any further study, there is a reasonable prospect of shortening the present rather lengthy HPLC runs, perhaps by having two isocratic steps, the first of which (for flushing out unwanted urinary constituents) might be done off-line.

Faster running might be acceptable, where the column resistance and the maximum pumping rate allow this. Use of a "wide" column (~4.4 rather than 2.2 mm i.d.) might increase resolution unless one uses a loop injector, but with the 4-fold greater cross-section sensitivity falls, the cost of packings rises, solvent costs likewise rise, and replenishment may be necessary after each run.

Acknowledgements
The work was supported by a grant from the Medical Research Council (to E. Reid). Professor J. H. Knox and colleagues kindly gave initial guidance on HPLC (to J. P. Leppard). Miss Barbara Brockway and Mr. R. J. Merritt gave excellent help, and Mr. A. A. A. Aziz collaborated in aspects such as setting up the reference method. Professor M. Sandler's group kindly advised on GC assay and donated phaeochromocytoma urines. The manufacturers of the packings have given sympathetic attention to the problems we have raised with them.

REFERENCES

1. Weil-Malherbe, H. (1971). *In* "Methods of Biochemical Analysis". Supplemental Vol. "Analysis of Biogenic Amines and their Related Enzymes". pp. 119-152. (D. Glick, ed.). Interscience, New York.
2. Aziz, A.A.A., Leppard, J.P. and Reid, E. (1976). *In* "Assay of Drugs and other Trace Organics in Biological Fluids". (Vol. 5 of "Methodological Developments in Biochemistry"). In press. (E. Reid, ed.). ASP Biological and Medical Press, Amsterdam.
3. Reid, E. (1976). *Analyst*, 110, 1-18.
4. Kahane, Z. and Vestergaard, P. (1967). *J. Lab. Clin. Med.* 70, 333-342.
5. Kahane, Z. and Vestergaard, P. (1969). *Clin. Chim. Acta.* 25, 453-458.
6. Smith, E.R.B. and Weil-Malherbe, H. (1962). *J. Lab. Clin. Med.* 60, 212-223.
7. Kappe, T. and Armstrong, M.D. (1965). *J. Med. Chem.* 8, 368-374.
8. Dell, D. (1976). As for Reference 2. In press.
9. Persson, B.-A. and Karger, B.L. (1974). *J. Chromatog. Sci.* 12, 521-528.

DETERMINATION OF ANTICONVULSANTS IN SERUM
BY USE OF HIGH PRESSURE LIQUID CHROMATOGRAPHY

Reginald F. Adams and Frank L. Vandemark

*The Perkin-Elmer Corporation,
Norwalk, Connecticutt, U.S.A.*

The potential of high pressure liquid chromatography for the routine analysis of drugs in physiological samples is just beginning to be explored. A procedure for serum theophylline has been reported (1). Recent reports (2,3) describe the separation of phenobarbital and diphenylhydantoin extracted from serum by use of an adsorption column with an organic solvent mobile phase. Here we used high pressure liquid chromatography with a reverse-phase column to determine phenobarbital, diphenylhydantoin, primidone, ethosuximide, and carbamazepine concentrations simultaneously in serum.

Materials and Methods

Apparatus

A high pressure liquid chromatograph (Perkin-Elmer Model LC 601) equipped with a reverse-phase column (ODS-Silex I) 0.5 m x 2.6 mm, was used together with a variable wavelength spectrophotometer (Perkin-Elmer Model LC 55) to monitor the column effluent. Data reduction was effected by use of the Perkin-Elmer PEP-II data processor. Special glassware included 16 x 75 mm disposable culture tubes, and 16 x 75 mm culture tubes (screw-capped with polytetrafluoroethylene liners).

Reagents

Acetonitrile (UV) distilled in glass. Burdick and Jackson Laboratories, Inc. Muskegon, MI 49442.
Mobile Phase: A solution of 85 ml acetonitrile (UV) in 415 ml deionized water. Both liquids should be degassed by

by cautious application of a vacuum, supplied by a water aspirator, immediately before gentle mixing.

Methanol: Spectroquality grade.

Charcoal: Norit A Neutral Amend Drug and Chemical Co., Irvington, N.J. 07111.

Mixed Solvent: Dichloromethane, isopropanol, diethyl ether 65/10/25 by volume. Spectroquality solvents must be used.

Drug Standards: We obtained carbamazepine (as Tegretol[R]) from Geigy Pharmaceuticals, Ardsley, N.Y. 10502. Methsuximide (as Celontin[R]) was obtained from Parke, Davis and Co., Detroit, Mi. 48232. Other drugs were obtained from Applied Science Laboratories, State College, Pa. 16801. 10 mg each phenobarbital, di-phenylhydantoin, primidone, methsuximide, carbamazepine, 50 mg ethosuximide were dissolved in 5 ml methanol and the volume exactly made up to 10 ml. The solution is stable at 4°C for at least two months.

Internal Standard: Dissolve 5 mg phenacetin (Applied Science Laboratories) in 10 ml methanol.

Procedure

Transfer 0.5 ml of serum, 2 ml of water, 10 µl [200 µl] of internal standard (equal to 5 µg of phenacetin) and about 4 mg of charcoal to a 16 x 75 mm tube. Mix the contents of the tube for 30 s with a vortex-type mixer. Centrifuge (2000 rpm, 1 min) and aspirate the aqueous phase. Add 300 µl of the mixed solvent to the tube. Using a vortex-type mixer, mix the tube contents vigorously for 30 s. Centrifuge (2000 rpm, 1 min) and decant the solvent into a 10 x 75 mm disposable culture tube. Evaporate the solvent in a gentle current of air (not greater than 50 ml/min at 40°C) to dryness. Take up residue in 20 µl methyl alcohol.

Aspirate 4 µl of the final volume into a syringe and inject directly onto the column. The chromatography is done routinely using the instrument conditions shown in Figure 1.

Results

Standard. To ascertain adequacy of the chromatography conditions aliquots of a standard mixture of the internal standard (phenacetin), ethosuximide, primidone, phenobarbital, methsuximide, diphenylhydantoin, and carbamazepine were applied to the column. The quantities injected were 1 µg each

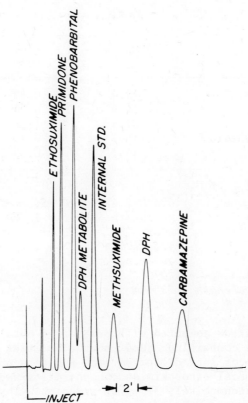

Fig. 1. Chromatogram of a standard mixture of drugs. Quantity chromatographed was 1 µg each except ethosuximide which was 5 µg. (DPH = diphenylhydantoin; DPH metabolite = 5 - (p-hydroxy-phenyl) 5-phenylhydantoin).

drug except ethosuximide which was 5 µg. The recorder was operated at 10 mV giving 0.2 absorbance units full scale. Figure 1 shows a chromatogram of a standard run. The analysis time from injection of the sample was about 14 min.

A mixture of drug-free serum and known quantities of added drugs was extracted and chromatographed. The drugs were identified by retention time. Similar chromatograms were obtained daily for reference use with extracted sera.

Sera. Sera from subjects on anticonvulsant therapy were extracted and chromatographed as described above. The resulting peaks were identified by retention time and measured

TABLE I

Drug	Mean* mg/l	S.D.	CV(%)
Ethosuximide	12.4	0.82	6.6
Primidone	14.6	0.61	4.2
Phenobarbital	31.4	1.50	4.8
Diphenylhydantoin	16.2	0.78	4.8
Carbamazepine	10.1	0.55	5.5

*n = 20

Column	0.5 M x 2.6 mm OD-Silex-I
Mobile Phase	Acetonitrile:water 17.83 parts by volume
Flow Rate	1.5 ml/min
Pressure	900 p.s.i.
Oven	40°C
Detector Wavelength	195 nm
Recorder	10 mV

using peak areas compared with peak areas obtained from extracted sera containing added known quantities of drugs. In the analysis of sera collected from subjects on ethosuximide therapy, typical ethosuximide concentration was found to be 68 mg/l; phenobarbital, 37 mg/l; and diphenylhydantoin, 31 mg/l.

Reproducibility

Data on within-run precision were obtained by taking 20 aliquots of each pooled sera known to contain primidone, phenobarbital and diphenylhydantoin. The aliquots were processed through the complete procedures during one day. The results are given in Table I.

Day-to-day precision data, collected for two other serum pools, are summarized in Table II. Duplicate aliquots of sera frozen until use of each of the pools were processed during each work-day over a four week period.

Background

When processed by our procedure, sera that were known

TABLE II

Drug	Mean* mg/l	S.D.	CV(%)
Ethosuximide	15.1	1.10	7.3
Primidone	11.8	0.74	6.3
Phenobarbital	35.9	1.90	5.3
Methsuximide	16.4	0.95	5.8
Diphenylhydantoin	14.7	1.00	6.8
Carbamazepine	9.8	0.72	7.4

*n = 20

to be drug-free yielded chromatograms showing varying background sometimes instrumental in origin. Background contamination was also contributed by glassware and solvents, and the rest comes from the sample itself and is related to the extraction procedure used. To obtain data on the amount of background appearing at elution time of the drugs of interest, we took 20 sera, known to be drug-free, with added internal standard (phenacetin) through the complete procedure. A range of values were calculated for background peaks appearing at the retention time intervals of ethosuximide, primidone, phenobarbital, methsuximide, diphenylhydantoin and carbamazepine. Aliquots of the same sera were processed without phenacetin but with added phenobarbital to provide an internal standard for the calculation of background in terms of phenacetin. The background concentrations that were negligible correspond to 0.01-0.06 mg/l.

Sensitivity

The sensitivity of the procedure is limited largely by serum volume, extraction efficiency, and background. With a 0.5 ml sample volume, sensitivity is adequate to permit detection of the compounds in concentrations as small as 0.1 mg/l except ethosuximide which was 0.5 mg/l.

Comparison with Gas Chromatography

We compared the procedures with gas chromatography as a comparison method. Additional aliquots of sixty-three of the same extracts used for the high pressure liquid chromatography were injected onto a 3% OV-17 column using gas chromatography conditions described elsewhere (4). Concentrations of phenobarbital and diphenylhydantoin were estimated from the peak areas compared with those obtained from extracted standards added to serum. Figure 2 shows the results plotted for phenobarbital. A comparison was carried out for diphenylhydantoin and primidone. The coefficients of correlation were calculated to be 0.95 and 0.96, respectively. The number of sera from subjects on therapy including ethosuximide, carbamazepine or methsuximide was not sufficient to carry out similar comparisons on those drugs.

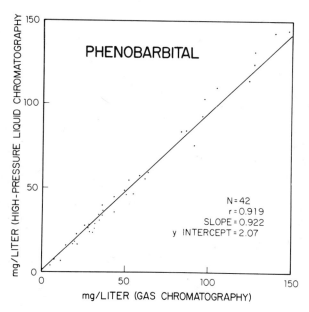

Fig. 2. Correlation plot showing comparison of results obtained from 63 sera by use of high pressure liquid chromatography and gas chromatography on aliquots of the same extracts.

Recoveries

We studied the recovery of drugs from serum by adding known quantities to pooled serum known to be drug-free. Half ml aliquots of the serum were taken through the extraction procedure. After cautious drying, the extracts were recon-

TABLE III

	Concentration mg/l	Mean Recovery Percent
Ethosuximide	1.0	42
	10.0	45
Primidone	1.0	65
	10.0	65
Phenobarbital	1.0	73
	10.0	72
Phenacetin	1.0	68
	10.0	71
Diphenylhydantoin	1.0	69
	10.0	71
Carbamazepine	1.0	58
	10.0	60
Methsuximide	1.0	44
	10.0	41

stituted in 20 µl of methanol containing 5 µg phenacetin. Aliquots of the solution were chromatographed and peak areas were computed. Since the phenacetin was not carried through the extraction procedure, it was used to correct for volumetric error in sample injection. Results were expressed as percentages of the peak areas obtained by direct injection of pure standards. Recoveries were calculated for phenacetin but using phenobarbital as the correction standard. Results were calculated for drugs recovered from duplicate serum samples containing drugs added to give concentrations equivalent to 1 mg/l and 10 mg/l. Table III gives the recoveries found and the mean concentrations for duplicate assays.

The recovery figures reflect the losses inherent in the micro-scale extraction procedure employed. The recovery may be improved by combining a second elution of the charcoal with the first. Adequate correction will be obtained by using standards consisting of sera to which the drugs have been added.

Quantitation

Quantitation was carried out by computer using standard chromatographic procedures based on peak area measurement. However, for small workloads, sufficient accuracy and speed may be obtained using peak-height ratio calculation. The internal standard, at a fixed concentration, is added to all samples. Sera containing known additions of drugs covering the required concentration range are processed and peak heights measured for each compound. Each measurement divided by peak height of the internal standard gives the peak height ratio for each concentration of the drug. A range of ratios is plotted and the plot serves as the calibration curve for subsequent unknown samples. An excellent exposition of this procedure is given in the report by Proellss and Lohmann (5). Although intended for gas chromatography, the principles also apply to liquid chromatography.

Drugs other than Anticonvulsants

To determine the potential usefulness of the procedures and to investigate possible interferences, several other ultra-violet absorbing drugs were studied by chromatographing pure standards. These included secobarbital, amylobarbital, salicylate, pentobarbital, methaqualone, chlordiazepoxide, glutethimide, propoxyphene. Diazepam and oxazepam did not elute from the column under the conditions used. Methaqualone alone eluted at 21.4 min and where present in a sample thus must be cleared before the next analysis. Amylobarbital emerged from the column with pentobarbital and methsuximide; secobarbital, glutethimide and propoxyphene with diphenylhydantoin. Salicylate (acetylsalicylic acid, salicylic acid) eluted immediately after the solvent peak and before ethosuximide, and thus did not interfere with the analysis. Retention times for the compounds tested are given in Table IV. No study of the drugs in physiological samples has been made.

Metabolites

The major metabolite of diphenylhydantoin 5-(p-hydroxyphenyl)-5-phenylhydatoin was studied. The recovery of the compound added to drug-free sera was 38-44%. The chromatography conditions used gave a retention time sufficient to

TABLE IV

Compound	Retention Time*
Salicylate	1.68
Ethosuximide	2.20
Primidone	2.93
Phenobarbital	3.98
5-(4-hydroxyphenyl)-5-phenyl	4.51
Phenacetin	5.51
Methsuximide	7.48
Pentobarbital	7.51
Amylobarbital	7.50
Secobarbital	10.15
Propoxyphene	10.15
Glutethimide	10.21
Diphenylhydantoin	10.25
Carbamazepine	13.40
Methaqualone	21.40
Oxazepam	Did not elute
Diazepam	Did not elute

*Retention Time measured from injection

allow resolution immediately following that of phenobarbital. The metabolite concentrations were calculated in 90 sera found to contain diphenylhydantoin. A range of values was obtained from 0.00-0.13 mg/l.

DISCUSSION

The above results indicate that high pressure liquid chromatography has adequate sensitivity and precision for quantitating the anticonvulsants tested in serum if most drugs are absent. The results obtained by comparison of gas chromatography and high pressure liquid chromatography on the same sera are excellent for primidone, phenobarbital and diphenylhydantoin. The analysis time and resolution of the chromatography using a reverse-phase column are related to the concentration of the organic solvent in the mobile phase. In practice, variation of the acetonitrile concentration provides considerable control in optimizing the analysis. The 17% concentration of acetonitrile was chosen to provide the shortest analysis time combined with resolution of the drugs required to be analyzed. The oven operating temperature, 40°C, was chosen to improve peak symmetry and to avoid variations associated with a varying ambient temperature.

The recoveries for the compounds studied are somewhat low because of the charcoal adsorption procedure. However, this is offset to a large extent by the convenience afforded by the small volume of solvent that is required.

A great advantage of high pressure liquid chromatography is that low concentrations of drugs may be analyzed without any requirement to derivatize the compounds. Carbamazepine may be quantitated in serum extracts without the precautions associated with gas chromatography procedures (5). The high pressure liquid chromatography detection system is non-destructive permitting the recovery of compounds for further examination.

The detection wavelength, 195 nm, although relatively non-selective, provides sufficient sensitivity for all the drug compounds tested. The lack of sensitivity is compensated for by the chromatographic separation of the drugs. Ultra-violet absorbing materials extractable from known drug-free sera have not presented a significant problem.

The number of samples that may be analyzed before column replacement depends on several factors. The most important of these will be the gradual loss of column efficiency because sample material carried over in the extract will coat the column packing. This will be evident as loss of resolution and shortened retention times. Also, the column back pressure eventually will increase until the ability of the pumping system to maintain the required flow is lost. The pressure increase is caused by particles from the sample or septum, clogging a porous plug sealing the end of the column. We routinely replaced the plug every 100 analyses. We have used one column for 600 analyses without significant loss of resolution. The chromatography time for each analysis was 14 minutes with an additional 6 minutes required for sample preparation.

Acknowledgement
We are indebted to J. Meola, Associate Director, Clinical Chemistry Department, Albany Medical Center, New York, for generous provision of sera. We thank W. Slavin for advice with this report and G. Schmidt for competent technical assistance.

REFERENCES

1. Weinberger, M. and Chidsey, C. (1975). *Clin. Chem.* 21, 834.
2. Evans, J.E. (1973). *Anal. Chem.* 45, 2428.
3. Atwell, S.H., Green, V.A. and Haney, D. (1975). *J. Pharm. Sci.* 64, 807.
4. Proelss, H.F. and Lohmann, H.J. (1971). *Clin. Chem.* 17, 222.
5. Perchalski, R.J. and Wilder, B.J. (1974). *Clin. Chem.* 20, 492.

ANALYSIS OF CARBAMAZEPINE IN PLASMA BY
HIGH PRESSURE LIQUID CHROMATOGRAPHY

I. M. House and D. J. Berry

Poisons Unit, New Cross Hospital, London.

The principle of monitoring antoconvulsant drug levels is well established, and because of its narrow therapeutic range and toxic effects the measurement of serum concentration is a useful ajunct to therapy with carbamazepine. This is a tricyclic compound with anticonvulsant properties and the drug of choice after phenytoin for control of grand mal epilepsy (1). The following is a quick, simple, precise and reliable method for its measurement.

Reagents

Diethyl ether, methanol and ammonium sulphate (Analar grade) were supplied by B.D.H. Chemicals, Poole, Dorset and imidostilbene by Ralph Emmanuel Ltd., Wembley, London. Carbamazepine (Analytically pure), 10,11-dihydrocarbamazepine, carbamazepine-10, 11-epoxide and 10,11-dihydroxycarbamazepine were kindly supplied by Geigy Pharmaceuticals, Hurdsfield Industrial Estate, Macclesfield, Cheshire.

Instrumentation

A liquid chromatograph (Applied Research Laboratories) incorporating a constant pressure pump, septum injection system and UV detector was used in conjunction with a potentiometric recorder.

Chromatographic Conditions

The column was a 300 mm x 2 mm I.D. stainless steel tube, packed by a balanced density slurry technique with Lichrosorb RP8 (Merck chemical obtained through B.D.H. Chemicals). The resolution obtained through this packing technique was good and reproducible.

Solvent:	Methanol: Water: 60: 40
Solvent Pressure:	100 bar (1400 psi) to give a flow rate of 1 ml min^{-1}
Detection Wavelength:	280 nm
Detector Sensitivity:	0.08 Absorbance units full scale

Extraction

The extraction was performed by adding 20 ml of ether to 1.0 ml of plasma in a stoppered, 35 ml centrifuge tube. This was shaken vigorously for 10 minutes, 2.5 g of ammonium sulphate was added and the contents of the tube shaken for a further 2 minutes. Use of this latter compound gave rise to a viscous plasma salt mixture which was easily separated from the organic phase by centrifugation at 3000 rpm for 5 minutes. The sulphate also performed the secondary function of drying the ether extract, which was poured into a clean 35 ml tube containing 1 ml of 10 µgml^{-1} of imidostilbene in ethanol. This solution was evaporated to dryness under a stream of air by transferral of 2-3 ml aliquots to a 10.0 ml conical tube partially immersed in a water bath at 40°C. The residue was dissolved in 100 µl of methanol and 5 µl of this was injected onto the column of the liquid chromatograph.

Quantitation

The chromatograph was calibrated by injection of a series of methanolic solutions containing known concentrations of carbamazepine and a fixed concentration of internal standard (imidostilbene) (Fig. 1). The peak heights were measured from an extrapolated baseline and the ratio of peak height drug to peak height of internal standard was rectilinear up to 10 µg in extract.

Recovery

The recovery of drug from plasmas at concentrations of 1, 2, 5 and 10 µgml^{-1} was assessed by adding 20, 40, 100 and 200 µg of carbamazepine respectively to 35 ml centrifuge tubes. This was performed by preparing a solution of 100 µgml^{-1} of carbamazepine in ethanol, pipetting the appropriate volumes into the tubes and evaporating off the solvent. 20 ml of citrated human blood bank plasma was added and the stop-

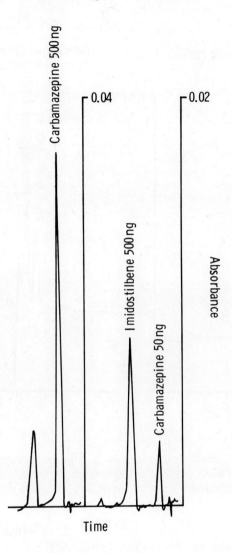

Fig. 1. Separation of Carbamazepine and Internal Standard under the chromatographic conditions specified.

pered tube shaken gently for a period of 4 hours. The solutions so obtained were treated as samples and analysed 10 times. The coefficient of variation was measured by analysing 56 samples in duplicate.

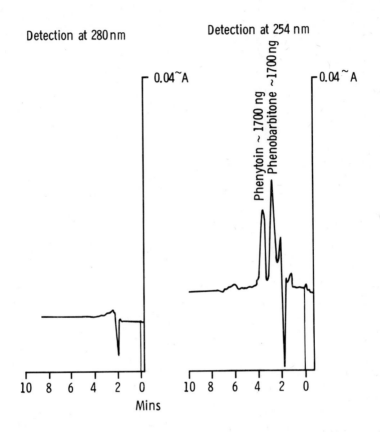

Fig. 2. Right hand trace shows a chromatogram of Phenytoin and Phenobarbitone with detection at 254 nm. Left hand trace shows a chromatogram of the same drugs with detection at 280 nm.

    Recovery   =   93 ± 9(S.D.) %
    Coefficient of variation   =   10%
    (analysis of duplicates)

Interferences

    No interferences have been observed either from the other common antiepileptic drugs (Fig. 2), or from endogenous plasma components.

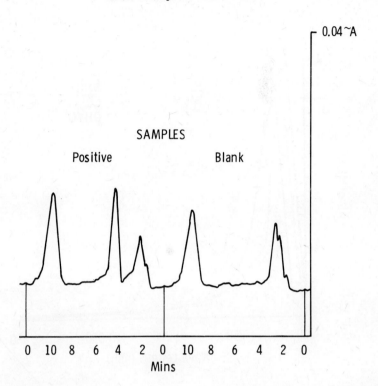

Fig. 3. Right hand trace shows a chromatogram of a plasma extract from a patient taking anticonvulsant drugs other than Carbamazepine. Left hand trace is from a patient on Carbamazepine therapy.

DISCUSSION

A column of partisil 10 (a 10 μ silica) was initially studied and although a wide variety of solvent mixtures was used, the resolution on this column was unsatisfactory since the carbamazepine peak was completely masked by interfering components extracted from plasma.

It was decided to evaluate a reverse-phase system, Lichrosorb RP8 (a 10 μ silica modified by chemically bonded C_8 n-alkyl groups). Figure 3 shows that this gives good resolution of carbamazepine (retention time 4 min) from extracted plasma components (retention time 2 min) and the internal standard (retention time 10 min), Imidostilbene. This last

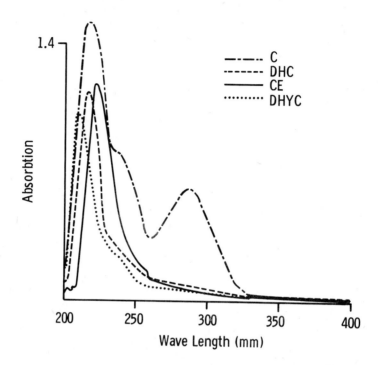

Fig. 4. UV Spectra of Carbamazepine and related compounds. C = Carbamazepine, DHC = 10,11-Dihydrocarbamazepine, CE = Carbamazepine 10,11-Epoxide, DHYC = 10,11-Dihydroxycarbamazepine.

compound is unstable and liable to slow photodegradation (hence the standard solutions are kept in the dark and prepared at monthly intervals). Increased selectivity was obtained by careful choice of detection wavelength (Fig. 2). 280 nm was chosen since this effectively removes any possible interferences by other antiepileptic drugs (since these have only low absorbance above 254 nm) and increases the sensitivity to carbamazepine (Fig. 4). However, the measurement of metabolites 10,11-dihydroxycarbamazepine and carbamazepine 10,11-epoxide is precluded since their absorbance at 280 nm is only $\frac{1}{100}$ of cabamazepine (Fig. 5).

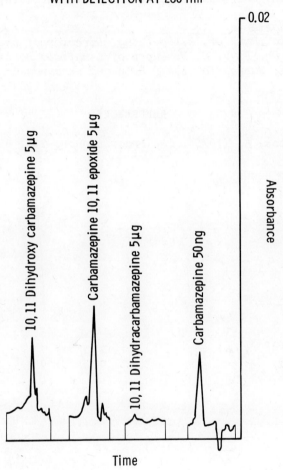

Fig. 5. Chromatograms of Carbamazepine and related compounds under the specific conditions.

Use

This method has been used to measure the carbamazepine concentrations in some 200 samples, including those from 56 institutionalised male adolescent patients. A scatter diagram of dosage versus plasma concentration for these results, (mean 4.1 µgml^{-1} with 95% range of 1.8 -9.1 µgml^{-1}) shows no significant correlation between the two parameters.

Acknowledgements
We wish to thank Dr. R. Goulding, Director of the Poisons Unit and Dr. P. V. C. W. Dupre of Lingfield Hospital School, Lingfield, Surrey for their help and encouragement with this work.

REFERENCES

1. Richens, A. (1975). *The Practitioner*, <u>215</u>, 653.

DRUG LEVELS IN PLASMA
SIMULTANEOUS DETERMINATION OF PHENOBARBITAL AND DIPHENYLHYDANTOIN

G. F. Johnston

Waters Associates (Instruments) Ltd.

ABSTRACT

Liquid chromatography has been shown to provide excellent quantitative separations of drugs from serum or plasma. Equally important where rapid analysis is imperative, liquid chromatography is a fast and simple technique.

A liquid chromatographic method is presented for the simultaneous determination of phenobarbital and diphenylhydantoin in plasma.

This work has been published in full by Atwell, S.H., Green, V.A. and Haney, W.G. (1975). *J. Pharmaceut. Sciences*, 64, 806-809.

THE HPLC DETECTION OF SOME DRUGS TAKEN IN OVERDOSE

P. F. Dixon and M. S. Stoll

*Department of Pathology,
Bromley Hospital, Kent.*

Emergency laboratories are commonly required to diagnose rapidly cases of drug overdose. In the Bromley Accident Centre barbiturate drugs are estimated as a group by a recording spectrophotometric method (1) with gas chromatography (2) for the estimation of the individual drugs. Since it should be the aim of the laboratory to confirm the diagnosis and to report a level of a specific drug as soon as possible, it was decided to investigate the use of HPLC for this analysis to see whether it offered any advantages over GLC in speed, ease or reliability.

The chromatography apparatus consisted of a Milton Roy single stroke pump, maximum pressure 1000 psi., connected via a septum injector to stainless steel columns (all purchased as individual items from Waters Associates); the detector was a Cecil Instruments CE 212 spectrophotometer fitted with an 8 µl flow cell of 1 cm path length. The operating wavelength of this instrument was continuously variable 200-450 nm; the output was recorded on a Perkin-Elmer flat bed recorder. The apparatus was laid out sequentially along a standard work bench and we feel that the easy accessibility of all the parts has contributed substantially to early success with the technique.

Our first separations of barbiturate drugs were carried out by adsorption chromatography on silica columns and are shown in Table I. Barbitone, phenobarbitone and quinalbarbitone were reasonably well separated but the others were too close together for practical purposes. Another disappointing feature of this system was the poor sensitivity, microgram amounts of each compound being required to obtain reasonably sized peaks. This was mainly because detection was at 254 nm, the wavelength of many fixed wavelength detectors, where unionised barbiturates absorb poorly. At shorter wavelengths

TABLE I

Separation of some Barbiturates in Silica Adsorption and C_{18}- Silica Bonded Reverse Phase Systems

	Corasil II Benzene/ Chloroform 50:50	C_{18} Corasil Methanol/ Water 60:40	C_{18} µ Porasil Methanol/ Water 60:40	C_{18} µ Porasil Methanol/ Water 50:50
	k'	k'	k'	k'
Barbitone	4.47	1.30	2.13	2.39
Phenobarbitone	3.67	1.45	2.57	3.48
Butobarbitone	3.13	1.70	3.04	5.17
Amylobarbitone	3.00	2.20	3.78	8.17
Pentobarbitone	2.93	2.20	3.78	8.17
Quinalbarbitone	2.33	2.67	4.30	10.70

with this system the solvent was opaque. We therefore turned our attention to reversed phase chromatography where aqueous methanol could be used as solvent. The separations which were obtained on C_{18} bonded to Corasil and µ Porasil are also shown in Table I. Although the separations of five of the drugs are good on C_{18} Corasil they are not complete unless the 10 µm particle size packing is used (Fig. 1). The marked increase of retention volume when the solvent polarity is only slightly lowered can be seen in Table I. The isomers amylobarbitone and pentobarbitone are not separated in any of the reversed phase systems.

Early experiments had apparently established that barbiturates absorbed maximally in aqueous methanol at about 216 nm and detection at this wavelength gave good sensitivity. But it was subsequently found that increased sensitivity was obtained at shorter wavelengths and was still increasing at 200 nm. This is shown for three of the barbiturates in Figure 2. A re-examination of the absorption spectra of the

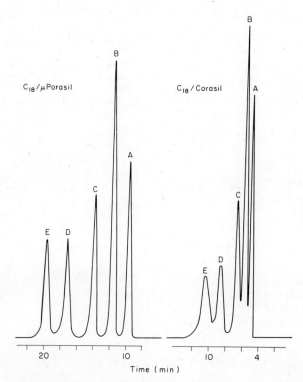

Fig. 1. Separation of barbiturates in a reversed phase partition system showing the effect of particle size. A: barbitone; B: phenobarbitone; C: butobarbitone; D: amylobarbitone; E: quinalbarbitone.

drugs showed that as they are diluted the maxima shift downwards until at the high dilution encountered in HPLC detectors they are below 200 nm. This might be accounted for by a breakdown of intramolecular associations with dilution.

A method using a single extraction procedure similar to that of the gas chromatographic method was proposed, with the following stages:
1. Extract 1 ml of serum with 2 ml chloroform by gentle rotation on a mixer for 5 minutes.
2. Extract 1 ml aqueous standard mixture and process in parallel for identification and quantitation.
3. Centrifuge and remove 1 ml lower layer.
4. Evaporate; redissolve in 0.1 ml 60% aqueous methanol.

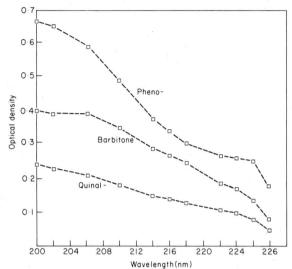

Fig. 2. The effect of detection wavelength on sensitivity for 1 µg amounts (injected) of phenobarbitone, barbitone and quinalbarbitone.

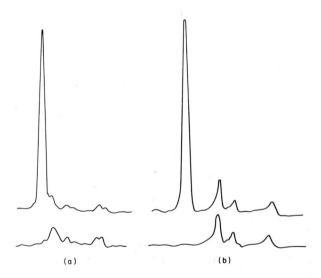

Fig. 3a. Chromatographs of two blank sera with and without phenobarbitone, 1 mg/100 ml (a) and butobarbitone, 1 mg/100 ml (b).

TABLE II

Maximum Variation of Triplicate Injections.
For Each Set of Triplicates the Difference Between the Highest and Lowest Value was Expressed as a Percentage of the Mean Value, and the Mean Percentage Variation was Calculated in the Four Optical Density Ranges

OD Range	No.	Mean Variation
.0009 - .01	15	3%
.01 - 0.1	15	2.1%
0.1 - 0.5	10	1.5%
> 0.5	6	2.4%

5. Inject 10 µl onto C_{18} µ Porasil column.

Chromatography was carried out using 60% aqueous methanol at 900 psi inlet pressure with detection at 200 nm.

Figure 3a shows an example of results which were achieved. Two different blank sera were extracted and chromatographed followed by the same sera with added amounts of 1 mg/100 ml of phenobarbitone and butobarbitone.

The main peak in the blank sera seems to be associated with extracted colour and may be bilirubin; it coincides with the barbitone peak. It can be reduced by an alkali wash procedure but since in practice barbitone overdose is rarely encountered it was decided not to include this step in the method.

The linearity and reproducibility of the chromatography have been tested by making triplicate injections of each of the five barbiturates in amounts from 10 ng to 4 µg. The response is linear up to 2 µg but in excess of this amount it flattens significantly in the cases of the more strongly absorbing compounds. The reproducibility of the injections was good and is shown in Table II. In many of the triplicates there was no measurable difference in the heights of the peaks. The sensitivity was sufficient to allow 5 ng of injected drug to be measured with reasonable accuracy.

A feature of HPLC is the ease and reproducibility with which injections can be made, and we feel that it is not necessary to use internal standards to compensate for injection variability. For example, when three separate extractions were made from the same patient's serum which contained pentobarbitone the heights of the peaks scarcely differed in

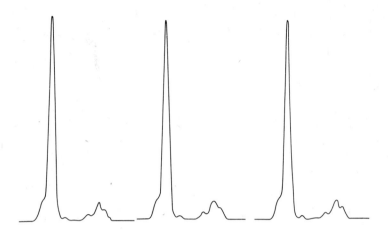

Fig. 3b. Chromatographs of three extracts of the same serum containing pentobarbitone, showing the reproducibility of the method.

TABLE III

Recovery of Barbiturates from Serum

Drug	% Recovery
Barbitone	68
Phenobarbitone	81
Butobarbitone	86
Amylobarbitone	*ca.* 100
Quinalbarbitone	90

size (Fig. 3b). The recovery of the five barbiturates is shown in Table III. Except for amylobarbitone a recovery correction is required as in the gas chromatographic method.

Although this method requires to be improved we feel that at this stage it can be compared with gas chromatography on a number of general points. Gas chromatography is superior in its ability to resolve completely all of the barbiturates and its sensitivity, using flame ionisation, is an order of magnitude greater than that of HPLC using UV detection. However, in this particular application sensitivity is not a problem. Under ideal conditions there is little to choose between the two methods for speed although our liquid chroma-

TABLE IV

Separation of some Tricyclic and Benzodiazepine Drugs in
Normal and Reversed Phase Silica Adsorption Systems

	Corasil II Methanol/ Water/ AqNH₃ 60:40:0.5	Corasil II Chloroform/ Ethanol/ AqNH₃ 200:20:0.1	Corasil II Iso:octane/ Isopropanol/ Ethanol/ AqNH₃ 90:4.5:5.5:0.5	Corasil II Iso:octane/ Chloroform/ Isopropanol/ Diethylamine 150:50:5:.005
	k'	k'	k'	k'
Carbamazepine	1.14	–	–	
Trimipramine	1.59	1.08	1.18	
Chlorimipramine	1.91	1.18	1.92	
Imipramine	2.16	1.25	2.48	
Nortriptyline	3.68	1.70	7.52	
Desimipramine	4.96	2.13	14.2	
Diazepan				1.93
Nitrazepan				3.93
Chlordiazepoxide				4.92

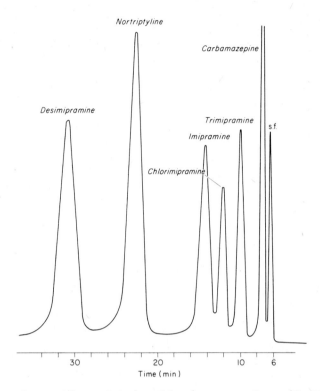

Fig. 4. *Separation of tricyclic drugs on Corasil II using methanol/water/aq. ammonia (s.g. 0.880), 30:20:0.25*

tography uses comparatively low pressures and could be speeded up. However, our experience is that HPLC nearly always performs ideally whereas gas chromatography is often fickle, so that in practice HPLC is nearly always considerably faster than gas chromatography. There is no doubt that with estimations starting from the cold machines, HPLC is superior since it is ready to use within minutes. We also feel that the operator skill required is considerably less than with gas chromatography. The technique seems to be more amenable to a general, as opposed to a specialist, operator than is gas chromatography.

Some separations of tricyclic antidepressant and benzodiazapine drugs are represented in Table IV. The tricyclic drugs can be separated in both normal and reverse phase adsorption systems but resolution is much greater in normal phase chromatography. They elute in two groups, the tertiary amines before the secondary amines. The related drug, carbamezepine, which is a primary amine is eluted ahead of all the others (Fig. 4). The order of elution is the same in both polar and non-polar systems (Table IV), with separation diminishing to zero at intermediate polarities. All systems need to be rendered basic in order to keep the amines in the non-ionised form.

We have only so far investigated the tricyclic drugs and the benzodiazapines as pure compounds and we have not extended these studies yet to extracts of biological fluids. It is certain that if blood levels are ever to be examined a very sensitive measuring system will have to be used since the tissues rapidly take up these drugs from the blood stream. In the systems just described a sensitivity of the order of 100 ng has been achieved and this is not high enough to permit blood levels to be determined; and it is likely that such an analysis would only be of value in confirming a suspected diagnosis, since the blood levels probably correlate poorly with the conscious state of the patient.

During the course of this work other workers have reported barbiturate analysis by HPLC (3,4,5), mainly using ion-exchanger systems and the chromatography of their dansyl derivatives with fluorimeter detection (6). Separations of twenty tricyclic drugs by ion-pair chromatography and adsorption chromatography on alumina have recently been reported in detail (7).

Acknowledgement
The authors are grateful to the South East Thames Regional Hospital Authority for a Locally Organised Research Grant with which the apparatus and materials were purchased.

REFERENCES

1. Broughton, P.M.G. (1956). *Biochem. J.* **63**, 207.
2. Flanagan, R.J. and Withers, G. (1972). *J. Clin. Path.* **25**, 899.
3. Ross, R.W. (1972). *J. Pharm. Sci.* **61**, 1979.
4. Evan, J.E. (1973). *Anal. Chem.* **45**, 2428.
5. Chan, M.L., Whetsall, C. and McChesney, J.D. (1974). *J. Chromatog. Sci.* **12**, 512.
6. Dünges, W., Naundorf, G. and Seiler, N. (1974). *J. Chromatog. Sci.* **12**, 655.

7. Knox, J.H. and Jurand, J. (1975). *J. Chromatog.* 103, 311.

APPLICATION OF HPLC TO THE STUDY OF THE DISPOSITION OF
HYDROCHLOROTHIAZIDE IN ADULTS AND CHILDREN

M. J. Cooper, M. W. Anders, A. R. Sinaiko and B. L. Mirkin

*Departments of Pharmacology and Pediatrics,
Division of Clinical Pharmacology,
University of Minnesota Health Sciences Center.*

The thiazides are among the most commonly used diuretic and antihypertensive agents. Since hydrochlorothiazide (6-chloro-3,4-dihydro-7-sulfamyl-2H-1,2,4-benzothiadiazine 1-1, dioxide) is much more potent than the parent compound chlorothiazide (6-chloro-7-sulfamyl-2H-1,2,4-benzothiadiazine 1-1, dioxide), it is administered in much lower dosages. Despite the use of these compounds over the past two decades, there is very little satisfactory data describing their disposition in the human. This has been due primarily to the existing assay procedures, which until recently were inadequate.

A new method has been developed for the determination of hydrochlorothiazide in serum and urine. It has been used to study the disposition of the drug in normal adult subjects, and also steady state serum levels in a random patient population.

Experimental

Analytical Procedures

The extraction and high pressure liquid chromatography was done according to Cooper *et al.* (1) and are shown in Figure 1.

Human Studies

Five healthy male volunteers, ranging in age from 22 to 47 years, were given a single oral dose of 1 mg/kg hydrochlorothiazide (Esidrix, Ciba Pharmaceuticals) following an overnight fast. In two cases (subjects 1 and 2, Figure 3), blood

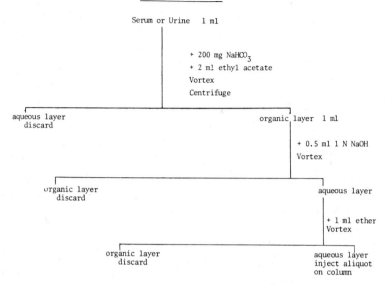

Fig. 1. *Extraction procedure and chromatographic conditions.*

samples were obtained by antecubital venipuncture before and 0.5, 1, 1.5, 2, 3, 4 and 5 hours after administration of the drug; urine samples were collected before and 1, 2, 4 and 6 hours after drug dosing. In the other cases (subjects 3, 4 and 5, Figure 3), both blood and urine specimens were collected prior to and at intervals of 1, 2, 4, 6, 8 and 10 hours after drug administration. All samples were stored at $-20°C$

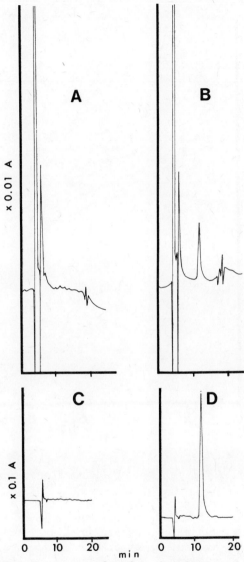

Fig. 2. HPLC of serum and urine samples.
A. Blank serum; B. Serum containing hydrochlorothiazide 800 ng/ml; C. Blank urine; D. Urine containing hydrochlorothiazide 80 µg/ml. 30 cm x 4 mm µ Bondapak C_{18} column with methanol-0.01 M sodium dihydrogen phosphate (1:4) as eluent; flow rate 0.6 ml/min; 271 nm detection.

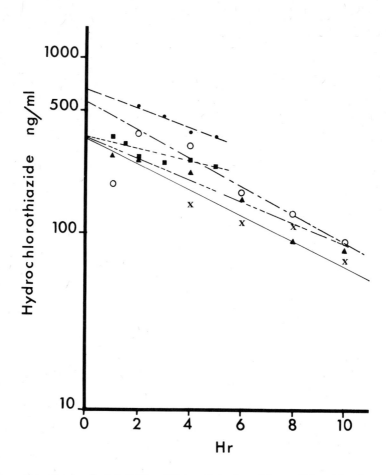

Fig. 3. Serum half-life (t½) of hydrochlorothiazide in normal subjects:
X ——————— X subject 1, t½ 4.4 hr
0 —·—·—·— 0 subject 2, t½ 3.7 hr
■ ------------- ■ subject 3, t½ 8.3 hr
● — — — — ● subject 4, t½ 5.2 hr
▲ —··—··— ▲ subject 5, t½ 4.6 hr

until analyzed.

Seven representative serum samples were obtained from a varied patient population from the University of Minnesota Hospitals. These subjects were on a steady state regimen of hydrochlorothiazide and several of them were being administered other drugs concurrently.

Analysis of Data

Serum half-life values were calculated by the method of least squares, using a single compartment model programmed for computer analysis.

RESULTS AND DISCUSSION

Methodology

The method is simple, rapid and sensitive. Representative chromatograms of serum and urine samples are shown in Figure 2. Recoveries are excellent; 98.2 ± 3.7% (mean ± S.E., n = 9) for serum and 91.5 ± 2.5% (mean ± S.E., n = 8) for urine (1). Moreover, many other drugs which may be administered concurrently during hydrochlorothiazide therapy were found to present no interference; included are azathioprine, chlorthalidone, guanethidine, methyldopa, minoxidil, prednisone and spironolactone. This is of special significance in the analysis of patient samples.

Serum Half-Life Values and Urinary Excretion of Hydrochlorothiazide in Normal Subjects Receiving a Single Dose of the Drug

The serum half-life values of hydrochlorothiazide in five normal male subjects receiving a single oral dose of 1 mg/kg are shown in Figure 3. The half-life was found to be 5.2 ± 0.8 hr (mean ± S.E.). The length of time following hydrochlorothiazide administration at which maximum serum concentrations were observed varied between individuals. In subjects 1 and 5, the peak was at 1 hr; in subjects 2 and 4, it was at 2 hr; while in subject 3, it was at 4 hr.

The maximum serum hydrochlorothiazide values found in this study were lower than those reported previously (2) and show considerable variability between subjects. This may be attributed to differences in drug absorption.

The cumulative urinary excretion data are shown in Figure 4. During a 6 hr collection period, 33.9 ± 3.5% (mean ± S.E.) of a single dose of hydrochlorothiazide was excreted in five subjects. After a 10 hr collection period in three sub-

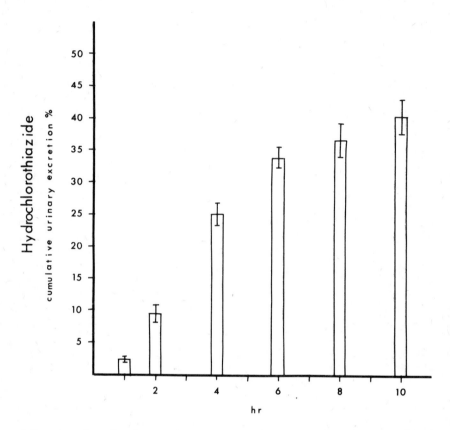

Fig. 4. Cumulative urinary excretion of hydrochlorothiazide in normal subjects.

jects, 40.6 ± 5.6% (mean ± S.E.) of the dose was recovered.

This urinary excretion pattern of hydrochlorothiazide in this study was similar to those previously published. Moyer et al. (3) have shown that maximum urinary excretion was reached within 3 hr in five normal subjects given a single oral dose of 1,500 mg hydrochlorothiazide. Anderson et al. (2) found in a study of four adults that 20-50% of a single oral dose of 65 mg of ^{14}C-labeled hydrochlorothiazide was excreted in 8 hr. In addition, McGilveray et al. (4) reported low levels of urinary excretion of hydrochlorothiazide in the first hour after ingestion.

TABLE I

Steady State Serum Levels of Hydrochlorothiazide in a Random Patient Population

Patient	Age Yrs.	Daily Dose (mg)	Serum Level (ng/ml)
D.D.	25	100	540
V.D.	13	125	132
A.G.	6	50	936
D.H.	24	200	2900
J.H.	15	100	450
C.N.	23	100	564
J.W.	19	50	1350

Steady State Serum Levels of Hydrochlorothiazide in Adult and Pediatric Patients

This study was undertaken as a preliminary step to demonstrate the utility of the method for clinical research. These patients were under treatment at the University of Minnesota Hospitals for a variety of pathophysiological conditions, and consequently were receiving a wide spectrum of drugs. The variation in serum concentrations observed (Table I) may reflect on the one hand the daily hydrochlorothiazide dosage, and on the other hand the degree of renal insufficiency. Since the accumulation of this drug may be a serious risk to the patient by enhancing the potential for hyperglycemia or hyperuricaemia, the significance of monitoring hydrochlorothiazide in serum should be investigated.

Acknowledgement
The authors thank Wilhelmina Ramos for technical assistance. This work was supported by funds from the United States Public Health Grant HD 08580, Program in Pediatric Clinical Pharmacology.

REFERENCES

1. Cooper, M.J., Sinaiko, A.R., Anders, M.W. and Mirkin, B.L. (1976). *Anal. Chem.* In press.
2. Anderson, K.V., Brettell, H.R. and Aikawa, J.K. (1961).

Arch. Int. Med. **107**, 168-174.
3. Moyer, J.Y., Fuchs, M., Irie, S. and Bodi, T. (1959). *Am. J. Cardiol.* **3**, 113-117.
4. McGilveray, I.J., Hossie, R.D. and Mattock, G.L. (1973). *Canad. J. Pharm. Sci.* **8**, 13-15.

AN ATTEMPT TO ISOLATE AND PURIFY DRUG METABOLITES BY HPLC FOR CHARACTERISATION BY PHYSICO-CHEMICAL METHODS

P. J. Simons

*I.C.I. Pharmaceutical Division,
Macclesfield, Cheshire.*

ABSTRACT

Limited data, based on the partial solution to 3 metabolism studies, is used to show that reverse phase HPLC has considerable potential for the isolation of small quantities of "clean" samples of radiolabelled metabolites for characterisation by GC/MS and NMR. Particular reference is made to detection, the problem of plasticisers and certain difficulties associated with the separation of closely related metabolites.

THE DETERMINATION OF THE SEMI-SYNTHETIC CEPHALOSPORIN
HR 580 IN PLASMA AND URINE

D. Dell

*Hoechst Pharmaceutical Research,
Walton Manor, Milton Keynes,
Buckinghamshire.*

INTRODUCTION

HR 580 is a semi-synthetic cephalosporin with the following structure:

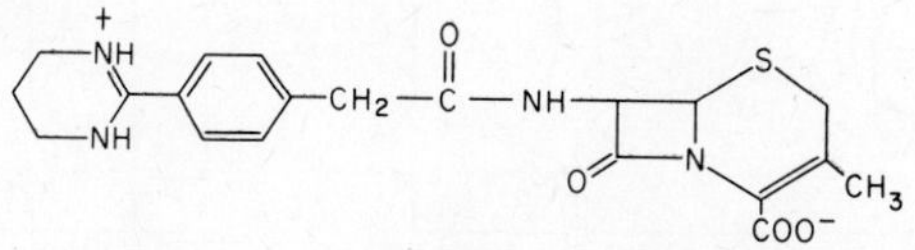

Microbiological methods are widely used for the analysis of penicillins and cephalosporins in biological fluids, and HR 580 is no exception. An alternative method of assay which would be more specific and at least as sensitive as the microbiological assay was required. As cephalosporins are relatively polar and non-volatile, a method involving high pressure liquid chromatography rather than gas liquid chromatography was investigated and developed.

Method

The detailed method which has been established in this laboratory for the determination of HR 580 in plasma or serum is as follows:
Internal standard (1 ml of a 16 mg/l aqueous solution of S 74 7534, an analogue of HR 580) is added to plasma (1 ml, in duplicate) and the resulting solution is freeze dried. Methanol (1 ml) is added to the dry residue and the mixture is vigorously agitated for 1 minute.
The sample is centrifuged at 2200 g for 3 minutes and the supernatant then transferred to a tapered glass tube.

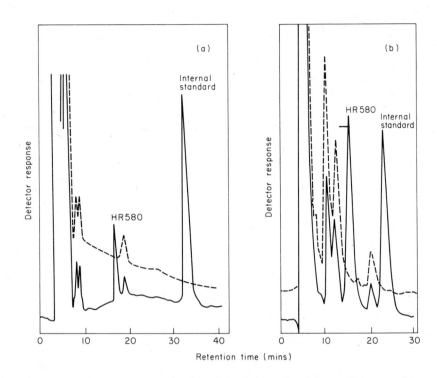

Fig. 1. Chromatograms of urine (a) and plasma (b) samples from a volunteer dosed with HR 580 (———) compared with control samples (- - -).

After concentration to about 100 µl in a stream of nitrogen, followed by centrifugation for 4 minutes, an aliquot (20-30 µl) of the supernatant is chromatographed as follows:

Waters M6000 liquid chromatography pump, U6K injector, and µ Bondapak C_{18} column (30 cm x 4.6 mm i.d.)

Mobile Phase: (a) 1% acetic acid/methanol, 83/17 v/v, for plasma samples;
(b) 0.1% Ammonium Carbonate/methanol, 75/25 v/v, for urine samples;
Flow Rate: 0.75 ml min^{-1} at 1500-2000 p.s.i.;
Detection: UV at 236 nm using a Cecil 212 monitor.

Under these conditions, the retention times are

(a) 16 and 24 minutes, and

(b) 17 and 33 minutes

for HR 580 and S 74 7534 respectively (Fig. 1). Peak heights are measured to ± 0.5 mm and peak height ratios are calculated (HR 580/S 74 7534).

Standard aqueous solutions of HR 580 are mixed with control plasma to obtain a series of samples in the concentration range 0.5 to 70 µg ml^{-1}. Internal standard, as described above, is added and the sample is taken through the same procedure. These samples are chromatographed and a standard curve is obtained by plotting peak height ratios against initial HR 580 concentration.

Urine samples are first passed through a 0.45 µ Millipore® membrane to remove suspended solids; they are then processed as described above with the exception that concentration by freeze drying is not necessary for all samples. Standard curves are obtained as for plasma, using control urine.

Evaluation of the Method

Sample Preparation

Because of the high water solubility of HR 580 (329 mg ml^{-1} at ambient temperature) and its low solubility in chloroform and n-butanol (1.0 x 10^{-3} and 4.2 x 10^{-3} mg ml^{-1} respectively) it was considered extremely unlikely that this compound would be amenable to conventional extraction procedures. Thus plasma and some urine samples had to be concentrated by freeze drying in order to attain final concentrations of HR 580 which could be detected and measured.

The importance of preserving the integrity of the top of a liquid chromatography column is now well established. Any mechanical disturbance of this area or accumulation of retained material quickly results in a drastic decrease in efficiency.

The aqueous alcohol mobile phase systems used routinely in reverse phase liquid chromatography can, if the proportion of alcohol in the mixture is high enough, provide suitable conditions for the precipitation of proteins when an untreated plasma sample is injected. It is essential, therefore, to remove proteins during the sample preparation step. After freeze drying the plasma sample, methanol is added; the denatured, insoluble protein may then be separated from the methanolic solution of the drug by centrifuging.

Failure to remove precipitated phosphates from urine samples either by centrifuging or Millipore®filtration, results in rapid deterioration in column efficiency. A column which has been thus poisoned may be regenerated by purging with a polar solvent such as dimethylformamide.

Choice of Column

Cephalosporins have been chromatographed on ion exchange columns such as Zipax SAX and SCX and AS-Pellionex-SAX resins (2). In general, efficiencies on ion exchangers are low (*ca.* 500 plates for 1 metre x 2 mm i.d.), and elevated temperatures (40-60°) have to be used to minimise mass transfer coefficients and thus maximise efficiency.

HR 580, being zwitterionic, can be chromatographed on both anion and cation exchangers using buffers over the pH range 4.0-7.0. Stability considerations precluded the use of extremes of pH. Because of the relatively poor efficiencies, limits of detection are moderate (150 ng injected gives 4% full scale deflection at 0.05 AUFS (absorption units full scale), signal/noise 10).

A μ Bondapak C_{18} reverse phase column was investigated. This consists of a monomolecular layer of octadecyltrichlorosilane permanently bonded to fully porous 10 μ silica particles via Si-C bonds.

The change from ion exchange to reverse phase brought about a marked improvement in the chromatography. Because of this increased efficiency (N = 1500 HETP = 0.2 mm), lower limits of detection were possible (10 ng injected gives 20% full scale deflection at 0.01 AUFS, signal/noise 10).

Mobile Phase

Water/methanol and water/acetonitrile are the most commonly used systems in reverse phase chromatography. Substituting 1% acetic acid or 0.1% ammonium carbonate for water sharpened the HR 580 peak.

For accurate quantitation, close control over the mobile phase composition is essential. In the example below, small increases in the proportion of the more polar component of the mobile phase mixture has resulted in a significant decrease in the relative retention times.

	Solvent Composition	Relative Retention (HR 580/Internal Standard)
Acetic acid/ methanol	80/20 83/17	0.68 0.58
Ammonium carbonate/ methanol	70/30 75/25	0.67 0.53

This, in turn, resulted in an increase in peak height ratio (HR 580/Internal Standard). Thus, any inadvertent small change in mobile phase composition will give rise to unreliable quantitation unless peak area, rather than peak height, ratios are being used.

Calibration

Calibration curves were obtained for each batch of analyses. Standard curves obtained over a period of 6 weeks show a variation of ± 5% about the mean line. The relationship between peak height ratio and HR 580 concentration was linear over the range studied (0.5-70 µg ml^{-1}).

Detection

The ultraviolet spectrum of HR 580 exhibits an absorption maximum of 235 nm in aqueous solution (E mol 2.16×10^4) and a shoulder at *ca.* 265 nm characteristic of the β-lactam moiety. The maximum at 235 nm is not shifted in acid solution (0.1 M HCl) although there is a hypsochromic shift in 0.1 M sodium hydroxide to 230 nm.

It is sometimes useful to sacrifice some sensitivity by selecting a wavelength which is 10 or 20 nm from the λ max for the compound in the event that the interference from endogenous material will be less at a different wavelength. This proved to be of no benefit in HR 580 assay.

Authenticity of HR 580 Peak

1. When added to urine or plasma samples, authentic HR 580 co-chromatographed with the peak in the sample attributed to HR 580 on a µ Bondapak column.
2. Fractions corresponding in retention times to HR 580 in urine samples were collected and analysed by TLC (sili-

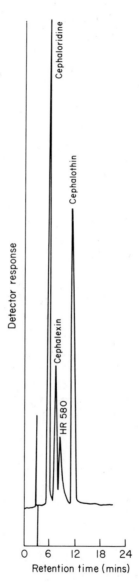

Fig. 2. The separation of cephalosporins on µ Bondapak C_{18}. Mobile Phase: 1% Acetic Acid:Methanol, 3:2.

ca) in 3 solvent systems:
   (a) Acetone:water, 70:30
   (b) N-butanol:water:96% ethanol:acetic acid, 50:20:15:15
   (c) Acetonitrile:water, 50:50

Three samples were run in each system:
   (i)   The collected fraction
   (ii) Authentic HR 580
   (iii) A mixture of (i) and (ii)

All three samples gave one spot only (visualisation by UV and iodine vapour) with identical Rf's.

3. The UV spectrum of the collected fraction corresponding to HR 580 was obtained and found to be identical to that of an authentic sample.

Specificity

A mixture of HR 580 with three of the most widely prescribed cephalosporins (Cephalexin, Cephaloridine and Cephalothin) was prepared in aqueous solution. Using a 60:40 1% acetic acid:methanol mixture, complete separation was achieved in 12 minutes (Fig. 2).

The need for specificity with respect to possible metabolites did not arise, because 100% of the dose was recovered in human urine as unchanged HR 580, as shown in the section describing application of the method.

Precision

Control plasma samples (10) were spiked with HR 580 to give a final concentration of 10 μg ml^{-1} exactly. These samples were taken through the established procedure and the following figures were obtained:

$$9.6 \pm 0.4 \text{ (SD)} \quad CV = 4.4\%$$

Precision was also evaluated from a number of duplicate results obtained from a clinical trial (Table I) by the method of Snedecor (3).

For values between 20 and 60 μg ml^{-1} SD (S) was found to be 0.2 μg ml^{-1} where N = 15; for values between 5 and 20 μg ml^{-1} S was 0.1 μg ml^{-1} (N = 10); for values between 1 and 5 μg ml^{-1} S was 0.1 μg ml^{-1} (N = 10).

Recovery

Control plasma samples were spiked with different

TABLE I

HR 580 Concentrations ($\mu g\ ml^{-1}$) in Human Serum by Microbiological Assay and High Pressure Liquid Chromatography, following an i.v. dose of 500 mg to Three Volunteers, A, B and C.

Sample Time After Dosing	A Microbiological	HPLC	B Microbiological	HPLC	C Microbiological	HPLC
5 Minutes	76.0		59.7	43.6	46.6	26.6
10 Minutes	64.7	51.0	50.8	36.0	39.8	22.4
15 Minutes	50.8	41.2	46.9	34.2	29.0	20.4
30 Minutes	46.9	33.3	34.0	27.8	24.8	15.5
45 Minutes	31.3	26.8	28.9	22.7	15.5	13.6
1 Hour	28.9	22.2	22.7	21.0	15.5	11.7
2 Hour	12.9	11.3	14.0	11.0	8.9	7.3
3 Hour	8.7	6.5	8.7	7.0	6.3	4.5
4 Hour	4.9	4.6	6.8	4.6	3.9	1.7
6 Hour	2.8	2.6	3.0	2.3	2.6	2.0
8 Hour	0	0.7	1.7	1.4	1.3	1.3
10 Hour	0	1.7	0	0.6	0	0

TABLE II

Excretion of HR 580 in Human Urine following an i.v. dose of 500 mg.

Subject	Collection Period (h)	Amount Excreted (mg)	% Dose HPLC	% Dose Microbiological Assay
A	0-6	455	91.0	110.5
B	0-6	453	90.7	101.0
	6-12	50.6	10.1	32.2
C	0-6	410	81.9	95.0

amounts of HR 580 to give ten samples over the concentration range 1-60 µg ml^{-1}. The samples were then taken through the method independently and the results compared with the "true" values:

HR 580 Concentration (µg ml^{-1}):

Added	Found	Difference as % Added
60.0	56.6	5.7
50.0	48.7	2.6
40.0	38.3	4.3
30.0	29.2	2.7
20.0	20.7	3.5
10.0	10.0	0.0
5.0	5.5	10.0
3.0	3.0	0.0
2.0	2.0	0.0
1.0	1.4	40.0

Mean difference from "true" result: 6.9%.

Sensitivity

This was discussed under *Choice of Column*. It was rarely possible to work at 0.01 AUFS with biological samples and the best that could be achieved for plasma samples was 0.5 µg ml^{-1}. This corresponds to about 150 ng injected.

Stability of Aqueous Standard Solutions of HR 580

At concentrations up to 20 mg/100 ml, aqueous solutions of HR 580 are stable (i.e. < 5% decomposition) at 0-4°C for up to two months as determined by HPLC.

Application

The method has been applied to a number of serum or plasma and urine samples from human volunteers and dogs and the results are collated in Tables I and II.

Acknowledgements
Blood and urine samples from human volunteers were provided by Dr. J. McEwen (Clinical Pharmacologist, Hoechst Pharmaceutical Research Ltd., Milton Keynes) and the Clinical Pharmacology Department of the West Middlesex Hospital in cooperation with the Medical Department of Hoechst (U.K.) Ltd.,

Hounslow. Blood and urine samples from a dog were provided by Dr. R. M. J. Ings (Pharmacokinetics Section, Hoechst Pharmaceutical Research Ltd., Milton Keynes). Mrs. J. Gibson (Analytical Services Department, Hoechst Pharmaceutical Research Ltd., Milton Keynes) ably assisted in the development and application of the HPLC method. Dr. Schrinner of Hoechst A.G. carried out the microbiological assays. Samples of HR 580 and synthetic analogues were kindly provided by Hoechst A.G., who also allowed this work to be published.

REFERENCES

1. Buhs, R.P., Maxim, T.E., Allen, N., Jacob, T.A. and Wolf, T.J. (1974). *J. Chromatog.* <u>99</u>, 609-618.
2. Cooper, M.J., Anders, M.W. and Mirkin, B.L. (1973). *Drug Metabolism and Disposition,* <u>1</u>, 659-662.
3. Snedecor, G.W. (1952). *Biometrics,* <u>8</u>, 85.

HPLC ANALYSIS OF CHLOROPHENOLIC POLLUTANTS AND OF THEIR OXIDATION PRODUCTS

D. C. Ayres and R. Gopalan

Department of Chemistry, Westfield College, University of London.

A separation of chlorophenols is described which is superior to existing g.l.c. procedures. Oxidation by ruthenium tetroxide is shown to destroy monochlorophenol pollutants and to raise the detection limits of those of higher chlorine number by converting them into quinones. The oxidation products which have been separated include those formed as secondary substances in the field and by metabolism. These pollutants are formed during the disinfection of industrial wastewaters by chlorination (1), whilst the use (2) of ^{36}Cl-labelled material led to the tentative identification in domestic waste of seventeen chloro-organic products including the 2,3, and 4-chlorophenols, 4-chlororesorcinol and 4-chloro-3-methyl phenol. The problem is exacerbated by the metabolism of chlorophenoxyacetic acids in water and in soil (3) and by the widespread use of impure highly chlorinated phenols as pesticides, a practice which is of concern to Government (4).

For the separation of phenols by gas chromatography derivitisation may be needed and detection limits by the standard methods are in the low p.p.m. range (5). Wolkoff and Larosse (6) reported the separation of several chlorophenols of different chlorine number by liquid chromatography. They used a method of detection based on the fluorescence of cerium (III) formed by the reduction of cerium (IV) solution, which was injected into the eluent stream. This substantially improved on the limits of UV detection at 254 nm but the response is progressively reduced as resistance to oxidation rises with increasing chlorine number: other practical difficulties of the technique have been discussed (7).

An effective separation of test groups of chlorophenols using a Porasil column, with a detection limit of tens of nanograms close to maximum UV extinction, is shown in Figure

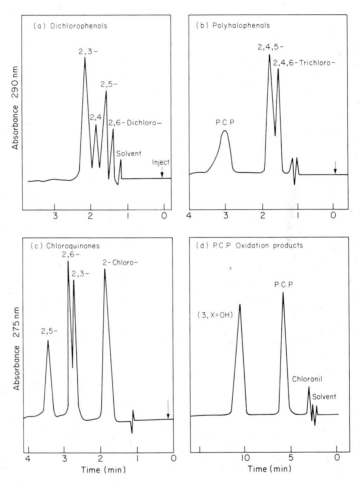

Fig. 1. HPLC Separations.
A: Dichlorophenols; B: Polyhalophenols; C: Chloroquinones;
D: P.C.P. Oxidation Products.

1a and 1b.

In the U.S.A. the levels of chlorinated derivatives in waste materials are high enough for the use of chlorine for disinfection to be questioned. We therefore examined an oxidative procedure for detoxification whereby the monochlorophenols are destroyed in neutral aqueous solution on contact

Scheme. *Products of Reaction between Chlorophenoxides and Ruthenium Tetroxide.*

Key to reaction conditions
A Ruthenium tetroxide in excess
B Phenoxide in excess
C Strongly alkaline medium

Ar = 2,4,6-trichlorophenyl
Ar' = pentachlorophenyl

with ruthenium tetroxide. This reagent may be generated in chlorine-water by the addition of catalytic amounts of ruthenium derivatives (8). Additional chloro-substituents reduce the rate of reaction (*cf.* the cerium (IV) method) and pentachlorophenol cannot be degraded in this way even after prolonged contact. However, the more highly chlorinated phenolate anions react rapidly in water at a pH of 8.0. With the oxidising agent in excess, chloroquinones are formed (Scheme, A routes, in *3* X=Cl) in yields ranging from 100% for chloranil down to *ca.* 50% for quinones of lower chlorine number which are themselves partially degraded. This reaction has potential for improving detection limits since the quinonoid oxidation products have molar extinctions (near 270 nm) an order of magnitude greater than those of the chlorophenols (near 290 nm). It should be noted that, in common with some other oxidising agents, the tetroxide displaces any *p*-chloro-substituent during quinone formation. Figure 1c shows the separation of all the possible chloroquinones that can be obtained from the di- and trichlorophenols previously resolved (1a and 1b).

When the 2,4,6-trichlorophenoxide ion is generated in excess by the addition of sufficient alkali to the phenol the principal products of the oxidation are those (*1* and *2* of the Scheme) formed by Michael addition (9) to the quinone with chlorine displacement. The analogous reaction occurs with sodium 2,4,5-trichlorophenoxide but not with the salt of pentachlorophenol, although the latter does form this kind of secondary pollutant in the field (10). The *in vitro* reaction of pentachlorophenoxide affords the dimer (*4*) as the sole product (11); it is probably formed from the mesomeric pentachlorophenoxy radical and it was characterised by m.p., mass spectrum (12) and the isolation of only pentachlorophenol on photolysis.

A clinical aspect is that induction of porphyria by hexachlorobenzene is accompanied by its metabolism to pentachlorophenol and other unknown oxidation products (13). Figure 1d illustrates the separation of compounds of this type including the extremely polar hydroxy derivative (*3*, X=OH) which is formed in the field (10). The dimer (*4*) is solvolysed during chromatography on Porasil and the detection of only pentachlorophenol in the eluent affords additional evidence of its structure. Polychlorobiphenyls have been linked (14) with porphyria and they are also known (15) to undergo hydroxylation *in vivo* to polychlorobiphenylols (*cf*. *5*). We have not so far extended our work on HPLC to compounds of this kind but we can report the high yield synthesis of 3,3',5,5'-tetrachloro-4,4'-dihydroxybiphenyl (16) (*5*) from 2,6-dichlorophenol when it reacts with ruthenium tetroxide in a strongly alkaline medium.

Experimental

The literature references 9, 11 and 16 include details of older oxidative reactions for the synthesis of products *1-5*, newly described here as products of ruthenium tetroxide oxidation. Full details of spectroscopic and analytical evidence of their identity will be given in a subsequent paper.

The HPLC separations were obtained with a Waters 6000 instrument using a Cecil detector and a 30 cm column of Porasil (10 μ, i.d. 6 mm) at ambient temperature and flow rates of *ca*. 1 ml/min. For separations 1a-1c we employed a solvent mixture of wet n-heptane (100 ml)-n-propanol (4 ml). For separation 1d a mixture of ethyl acetate (115 ml):n-heptane (75 ml):m thanol (10 ml) was used.

Acknowledgement
We thank Dr. C. K. Lim for his skilled assistance and advice with the HPLC.

REFERENCES

1. Murphy, K.L., Zaloum, R. and Fulford, D. (1975). *Water Research*, 9, 389.
2. Jolley, R.L. (1975). *J. Wat. Pollut. Control Fed.* 47, 601.
3. Allebone, J.E., Hamilton, R.J. and Ravenscroft, B. (1975) *Environmental Chem.* 1, 160.
4. "The Non-Agricultural Uses of Pesticides in Great Britain". 1974. H.M.S.O.
5. Fountaine, J.E., Joshipura, P.B., Keliher, P.N. and Johnson, J.D. (1974). *Anal. Chem.* 46, 62.
6. Wolkoff, A.W. and Larosse, R.H. (1974). *J. Chromatog.* 99, 731.
7. Katz, S. and Pitt, W.W. (1975). *J. Chromatog.* 111, 470. See also a reply in this Journal, p. 472.
8. Ayres, D.C. and Hossain, A.M.M. (1975). *J. Chem. Soc. Perkin*, 1, 707.
9. Hunter, W.H. and Morse, H. (1926). *J. Amer. Chem. Soc.* 48, 1615.
10. Kuwahara, M., Kato, N. and Munakata, K. (1966). *Agr. Biol. Chem. (Japan)*, 30, 232.
11. Reed, R. (1958). *J. Amer. Chem. Soc.* 80, 219.
12. This was determined by the P.C.M.U. at Harwell and we are indebted to Professor A.H. Jackson for a confirmatory field desorption spectrum.
13. Lui, H. and Sweeney, G.D. (1975). At the 1st International Porphyrin Meeting, "Porphyrins in Human Disease", Freiburg.
14. Elder, G.H., Evans, J.O. and Matlin, S. Reference 13 contribution.
15. Zitko, V., Hutzinger, O. and Choi, P.M.K. (1975). *Bull. Environ. Contamination and Toxicol.* 13, 649.
16. Inoue, H., Simamura, O. and Takamizawa, K. (1962). *Bull. Chem. Soc. Japan*, 35, 1958.

AN EVALUATION OF SOME HPLC COLUMNS FOR THE IDENTIFICATION AND QUANTITATION OF DRUGS AND METABOLITES

P. J. Twitchett, A. E. P. Gorvin, A. C. Moffat,
P. L. Williams and A. T. Sullivan

*Home Office Central Research Establishment,
Aldermaston, Reading, Berks.*

The first part of this paper is devoted to our evaluations of some of the available HPLC separation modes used in drug analysis. This will be followed by examples which illustrate the way in which HPLC is being used at the Home Office Central Research Establishment for the analysis of drugs in biological materials. The illustrations will show what are considered to be the particular advantages in this work:

The ability of HPLC to handle aqueous samples directly, sometimes in millilitre volumes, without the need for preliminary extraction procedures.

The use of reverse-phase separation systems where elution is in order of polarity, and the more polar drug metabolites are eluted before the parent drug.

The use of specific detectors to enable the detection of trace quantities of drugs or metabolites against a large background of other components, for example, in urine.

The simple preparative use of HPLC, allowing the separated components of a mixture to be collected for further analysis.

Evaluation of HPLC Systems for Drug Analysis

HPLC offers several modes of separation, and a profusion of column packings. As many drug substances comprise both ionic and non-polar functional groups it is not surprising that similar separations may be achieved by more than one chromatographic mode. A brief survey of the literature reveals that drug mixtures containing diamorphine, for example, have been separated with varying degrees of success by high speed anion - and cation-exchange, liquid-solid adsorption and normal and reversed-phase liquid-liquid partition chromatography. Thus it is not at all clear where to begin in choosing the optimum chromatographic medium and conditions for a particular separation problem. This difficulty is particularly acute in the field of forensic science, where the analyst may be required to deal with aspirin one day and illicit amphetamine or the weedkiller paraquat the next. Hence there appeared to be a need for a rational investigation of the various chromatographic modes and columns available to determine their suitability for the analysis of different classes of drug. Part of our work at the H.O.C.R.E. has been in this direction.

For these studies, thirty compounds were selected as representative of a wide variety of polar drug substances, strongly acidic, weakly acidic, neutral and basic, and of differing chemical structure, molecular weight, lipid solubility and pharmacological action. From a knowledge of the chromatographic characteristics of these compounds, the behaviour of many other drugs may be predicted.

Octadecylsilane (ODS) bonded phase columns have a wide applicability and it was to these that we first turned our attention (1). A number of inter-related parameters are important in the operation of a reverse-phase system, the eluent pH and organic solvent content must be matched against the pKa and lipid solubility of the drug. The almost exponential increase in retention volume (Rv) with decreasing eluent methanol content is well known to those familiar with reverse-phase chromatography, but a good example of a less well appreciated although equally predictable variable is shown by the dependence of retention volume with eluent pH. For acidic drugs, there is a marked decrease in Rv as the eluent pH is increased above the drug pKa, while for bases, the Rv increases with increasing eluent pH. Evidently retention on an ODS column is directly related to the degree of ionisation as well as the lipid solubility of the drug. Lipid solubilities are commonly measured in terms of the n-octanol/water partition coefficients (P), and there is a reasonable correlation between these partition coefficients and retention volumes measured at an appropriate eluent pH (Table I). Hence eluent pH is another useful parameter that may be varied most predictably to optimise separations of ionisable drugs according to their pKa. However, although ODS columns may be used for the chromatography of almost any type of drug, any generally useful column must have a reasonable

TABLE I

Drug	pK_a	log P	Retention Volume (ml) (Eluent pH = 3)
Barbitone	7.8	0.7	3.7
Phenobarbitone	7.4	1.4	4.4
Quinalbarbitone	7.9	2.3	6.9
Thiopentone	7.6	3.0	7.8

P = n-octanol/water partition coefficient.

chromatographic efficiency, as this governs the resolution and sensitivity that may be achieved. In this respect we found that whereas ODS gave excellent results for acidic drugs, neutral compounds and diazepines, the column efficiency for bases was sometimes very low indeed.

Previous experience with Zipax SCX, a pellicular cation exchange material, had indicated that cation exchange chromatography could be valuable for the separation of basic drugs, which by their nature are well suited to ion-exchange procedures. It was hoped that such a system would complement the ODS column for basic drugs.

For the evaluation of a recently available microparticulate cation-exchange column (Partisil-SCX) the same thirty drugs were again used. Acidic and neutral compounds were included as preliminary experiments had indicated that non-ionic interactions could give rise to some retention even for non-basic compounds. Chromatographic retention and efficiency were measured using aqueous eluents of varying ionic strength, pH and organic solvent content. The results indicated that retention was dependent on two major factors:

 1. Cation exchange - influenced by varying the eluent ionic strength.
 2. Reverse-phase partition - influenced by variation of the eluent pH and organic solvent content.

The dependence of Rv on eluent ionic strength {I} proved to be a simple relationship, Rv increased in inverse proportion to {I}. The magnitude of the effect was proportional to the drug basicity, as expected for a cation exchange column. The variation of eluent pH however, produced results contrary to those expected for a cation exchanger, but both the pH effect and the effect of added organic solvent in the eluent followed closely the behaviour of a reverse-phase partition column. Measurements of Rv and column efficiency showed that for many drugs an eluent containing an organic solvent was necessary, not only to elute substances from the column but also to enhance the column efficiency which was poor only if eluents containing less than 40% methanol or acetonitrile were used.

The cation exchange column has therefore proved a reasonable chromatographic medium with a readily predictable mode of action based on both cation exchange and partition chromatography. Not surprisingly, although some retention was observed for non-basic drugs there was little selectivity. Unfortunately, though, the value of the column was diminished by the short life of the columns which we have observed, especially at eluent pH values greater than 7 (2).

At the Central Research Establishment, HPLC is used to deal with three types of problem with which we are concerned.
 1. Separation and identification of the components of illicit drug mixtures.
 2. Detection of drugs in biological fluids.
 3. Investigation of certain aspects of drug metabolism.

In each field of work there is the problem of separation of complex mixtures, although the detector sensitivity required becomes much greater with the biological and metabolic samples.

Illicit drugs are frequently complex mixtures comprising degradation products, synthetic precursors and various additives and diluents, as well as the active principle. A sample of illicit "Chinese heroin" was recently sent for analysis to C.R.E. The case was unusual in that analysis by gas-chromatography/mass spectrometry (g.c./m.s.) indicated the presence of an unusual alkaloid thebacon. Examination of

Acetylcodeine   *Thebacon*

the sample by HPLC using cation exchange and silica columns showed that no thebacon was present, but rather a substance subsequently identified as acetylcodeine. The two compounds are isomeric and differ only in the position of a C=C double bond, thus giving mass-spectral fragmentation patterns which were almost identical. This is an example of the combined use of high resolution liquid chromatography with mass spectrometry in a case where m.s. alone would give inconclusive results.

The detection of drugs in biological samples is an area in which HPLC has many advantages, not least of which is the rapidity with which samples may be analysed without prior extraction or clean-up. A very simple example is a recent case involving suspected aspirin poisoning which illustrates the use of the octadecylsilane column for the analysis of acidic drugs. The victim was a child, and the circumstances suggested that death was due either to self-poisoning with "junior aspirin" tablets (which are orange in colour) or to the deliberate administration of adult aspirin tablets with orange juice. "Junior aspirins" contain both aspirin and saccharin, and orange juice contains saccharin and benzoic acid. It was necessary to compare the amounts of salicylic and acetylsalicylic acid, saccharin and benzoic acid present in the stomach contents of the dead child with the amounts in "junior aspirin" and orange juice. By the use of reverse-phase HPLC these four components together with citric and ascorbic acids could be separated in less than four minutes. No preliminary extraction procedures were required as samples of both orange juice and stomach contents could be injected directly onto the column.

The hallucinogen lysergic acid diethylamide (LSD) is difficult to detect in body fluid samples, as the effective dose is only 50-100 μg and the metabolism is not well understood. However, LSD itself is highly fluorescent and thus amenable to analysis by HPLC with fluorimetric detection, where the sensitivity is in the picogram region. In the forensic science context, it is necessary not only to analyse drugs quantitatively, but also to prove conclusively the identity of the drug, and unfortunately, drug-takers do not oblige the analyst by providing control urine samples. In attempts to assess and enhance the specificity of current radioimmunoassay (RIA) methods for LSD, several case samples have been examined concurrently by radioimmunoassay and by HPLC. Using the HPLC method, the analysis of stomach washings from a patient who subsequently died, showed a level of 6 ng LSD ml^{-1}. The analysis was performed by the direct injection of sample onto an ODS column, and by means of a valve arrangement the chromatographic peak with a retention volume corresponding to LSD could be trapped in the detector flowcell. Using a scanning fluorimeter as an HPLC detector both the excitation and emission spectra of the trapped component could be recorded to show that the chromatographic peak had fluorescence characteristics as well as a retention volume identical with LSD. Furthermore, after irradiating the sample with 320 nm radiation in the flowcell for a few minutes, the fluorescence intensity was reduced. Irradiation of aqueous LSD solutions leads to the formation of the non-fluorescent lumi-LSD by photo-addition of water. In this case a radioimmunoassay gave an identical level of LSD, whereas urine samples giving positive RIA analyses showed no detectable LSD by the HPLC method. The RIA method cross-reacts with metabolites and little unchanged LSD is present in urine.

In the study of drug metabolism, a simple experiment can demonstrate some of the advantages of HPLC. Aspirin metabolites for example may be readily detected in the urine over 24 hours after the ingestion of a single dose (600 mg) of aspirin. This is possible by the direct injection of only 5 µl of neat urine onto an ODS column. Using U.V. absorbance, detection of the drug and metabolites amongst the very many U.V. absorbing acidic urinary components is not possible, but with the more sensitive and selective fluorimetric detector, the metabolite salicyluric acid may be readily observed (Fig. 1). Again, the ability of HPLC using a specific detector to deal with urine samples directly and without prior extraction leads to a rapid and sensitive analysis.

Cannabis is another drug for which the methods of detection in body fluids are still under development, and for this HPLC is being used as a separation method prior to a radioimmunoassay procedure. An immunoassay method for tetrahydrocannabinol (THC) has been developed at the University of Surrey. Initial results on blood and urine samples from cannabis smokers indicated higher levels of cross-reacting material than could be accounted for by the presence of THC, suggesting that the assay was reacting to metabolites of THC. By using liquid chromatography to fractionate urine samples and separate the metabolites, it is hoped to discover which metabolites the assay is detecting. For this purpose,

Detection of Aspirin Metabolite (Salicyluric Acid) in Urine

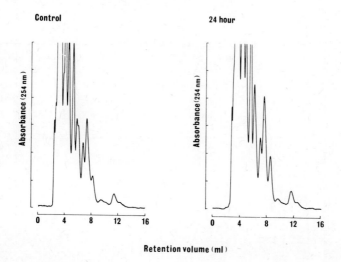

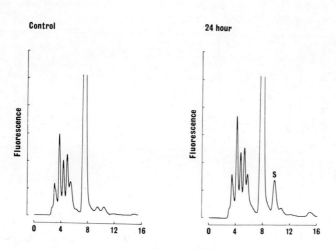

Fig. 1. Detection of an aspirin metabolite in 5 µl of urine. S = salicyluric acid. Column: octadecylsilane bonded to Partisil -5. Eluent: pH 3.0 buffer containing 20% methanol. Upper chromatograms: u.v. detector (254 nm). Lower chromatograms: fluorimetric detector λex 295 nm λem 400 nm.

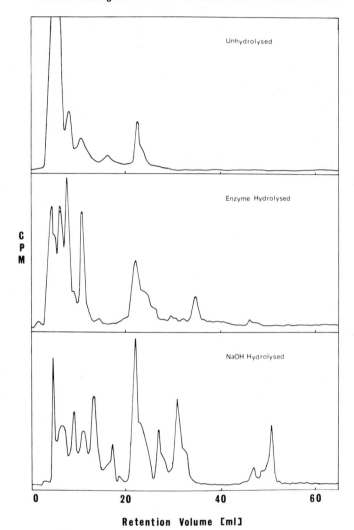

Fig. 2. Radiochromatogram of tetrahydro - cannabinol metabolites in rabbit urine after hydrolysis by various methods. Column: octadecylsilane bonded to Partisil -5. Eluent: 50% aqueous methanol step gradient at 20 min (65% aqueous methanol) step gradient at 40 min (80% aqueous methanol). Detection by fraction collecting and scintillation counting of 14_C activity.

and HPLC detector of greater sensitivity than the fluorimeter was required. ^{14}C labelled THC was therefore administered to a rabbit, and urine samples collected over a 3 day period were chromatographed. 0.5 ml fractions of eluent were collected and subsequently counted using a scintillation counter. By the use of a reverse-phase column (ODS) up to 2 ml quantities of urine could be injected and analysed; metabolites were then eluted in the order of polarity. As the nature of the metabolites was not known, it was important to ensure that all THC-derived substances were eluted. On a reverse-phase system such as this, however, the elution conditions may be set up using the parent drug as the more polar metabolites will have retention volumes less than that of the drug.

A radiochromatogram of an untreated urine sample was relatively simple (Fig. 2). The very high polarities of the major components were indicative of conjugates, and indeed, treatment of the urine with β-glucuronidase and sulphatase led to some changes in the chromatogram, but with methanolic 0.2M NaOH even greater hydrolysis was obtained. Surprisingly, some of the known metabolites of THC were stable under these hydrolysis conditions. The value of liquid chromatography methods in work of this sort is considerable, hydrolysis procedures can be rapidly optimised, the method has a greater resolution than thin layer chromatography, and the use of reverse-phase columns ensures that no very polar or thermally labile metabolites are lost. The eluted metabolites are now being used in the RIA procedure and for mass-spectral analysis. It is hoped that a combination of these three powerful techniques will lead to a reliable method for detecting in body fluids this most common drug of abuse.

REFERENCES

1. Twitchett, P.J. and Moffat, A.C. (1975). *J. Chromatogr.* **111**, 149.
2. Twitchett, P.J., Gorvin, A.E.P. and Moffat, A.C. (1975). In press.

SEPARATION OF DRUGS BY HPLC AND THE APPLICATION OF FLUORIMETRIC DETECTION TO DRUG PROBLEMS

B. B. Wheals

Metropolitan Police Forensic Science Laboratory, London.

The analytical problems occurring in the drug section of a forensic science laboratory can be very different from those which a typical pharmaceutical laboratory might encounter. In general the analyst has no control over the type of sample with which he is presented and often has little or no indication of its contents. Sample size can be very variable, ranging from kg amounts of materials (e.g. large seizures of cannabis) down to µg or ng amounts of drugs if a syringe washing or a toxicological sample requires analysis. Forensic work also demands that the identities of any drugs encountered be established beyond reasonable doubt - hence qualitative analysis generally takes precedence over quantitative. For rapid qualitative analysis, however, thin-layer chromatography using a variety of spray reagents is markedly superior to high pressure liquid chromatography (HPLC) in terms of speed and low cost, and one might be tempted to predict that HPLC could only play a very minor role in forensic drug analysis. Experience we have built up over the last few years, however, has shown that HPLC can make a very useful contribution, particularly when mixtures of illicitly prepared drugs are being analysed. A review of these various applications has been published elsewhere (1).

For the analysis of low levels of materials the use of fluorimetric detection in combination with HPLC offers a powerful technique, although it has not been exploited to the same extent as UV detection. Under optimum instrumental conditions fluorimetric detectors are capable of giving a 100-1000 fold increase in sensitivity (i.e. with detection limits in the region of 10-100 pg) over their UV counterparts. Like UV detectors the fluorimetric detector is selective (by tuning of excitation and emission conditions), but the latter

TABLE I

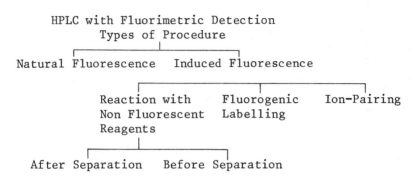

has the advantage of being unaffected by solvent pulsing. There are now fluorimetric detectors available commercially and many fluorimeters can be easily modified to become detectors by introducing a suitable flow cell. In our experience it is not easy to develop a flow cell with a capacity of less than $ca.$ 50 µl, for if right angled optics are being used appreciable light scatter occurs with silica tubing of less than about 2 mm i.d., and it is necessary to have at least 10 mm of the tube illuminated to obtain acceptable signals from the sample.

In applying fluorimetric detection to drug problems it is obviously essential that the compound being analysed display photoluminescence properties. There are relatively few drugs of abuse which have natural fluorescence but by various routes it is possible to chemically induce this property (see Table I). It is unfortunate that many condensation reactions leading to the formation of fluorescent products require concentrated sulphuric acid, for a less desirable medium for subsequent chromatography is hard to imagine. In our laboratory we have sought to exploit far less rigorous conditions.

Compounds Detected by Natural Fluorescence

Amongst drugs of abuse, lysergic acid diethylamide (LSD) is one of the few compounds to display very high levels of natural fluorescence λex 320 nm λem 420 nm. The high activity and therefore low active dosage of this drug (i.e. $ca.$ 100-150 µg/trip) necessitates the use of a sensitive analytical technique for its detection. Illicitly prepared microdots can be very rapidly screened for the presence of LSD by extracting with methanol and injecting an aliquot of the resulting solution on to a silica HPLC column coupled to a fluorimetric detector (2,3). In a 5-10 min analysis LSD can be separated from related ergot alkaloids (which are also fluorescent) and quantitatively detected. Under the chromatographic conditions used, LSD has a characteristic retention time not shown by any other compound of the same fluorescent properties that we have tested. UV absorbing materials which are present in the tablet extracts, but show no fluorescence, go undetected. As little as 10 pg of LSD can be detected by the procedure and it has been possible to detect LSD in the urine of persons taking the drug (4).

Compounds Detected by Induced Fluorescence

Reaction with Non-Fluorescent Reagents before Chromatography

Reaction before chromatography can be a powerful method of inducing fluorescence but this must not lead to the introduction of materials on to the column which will interfere with the subsequent separation.

A typical example of this type of approach is a method developed for the analysis of morphine in body fluids using HPLC with fluorimetric detection (5). Morphine can be converted to a fluorescent dimer - pseudomorphine, by mild oxidation, but morphine analogues also produce fluorescent dimers under the same conditions so the reaction is not specific. It is, however, possible to separate these dimers by HPLC. In the method used, the oxidative dimerisation is performed on the top of a silica column, followed by elution and detection of the pseudomorphine. The presence of co-extractives was found to cause a variable reaction yield but this problem was overcome by introducing a reactive internal standard - dihydromorphine. This compound also undergoes the dimerisation reaction and when it is present with morphine three oxidation products are produced (i.e. morphine dimer, dihydromorphine dimer and morphine-dihydromorphine). The three products are separated under the chromatographic conditions used and from the relative peak heights it is possible to determine the morphine concentration in the sample.

Reaction with Non-Fluorescent Reagents after Chromatography

If some form of reactor is introduced after the chromatographic column it is possible to use drastic reaction comditions to produce fluorescent products from the eluting compounds. The major problem is one of keeping the dead volume of such a system sufficiently low so that band spreading does not lead to excessive peak overlap. At the present time we are not using a method of this type of forensic work but some excellent results have been obtained by Katz and co-workers (6) in the bio-medical field. In their procedure a reactant solution of 10^{-4}N Ce^{4+} in 2 N sulphuric acid is mixed with the eluting solvent stream. With suitable compounds oxidation occurs to form Ce^{3+} which displays fluorescence (λex 260 nm λem 350 nm). The method has been used to detect dicarboxylic acids in urine samples following ion-exchange chromatography - detection limits 100- 500 ng.

Fluorigenic Labelling

Although fluorigenic labelling can be classified under *Reaction with Non-Fluorescent Reagents before Chromatography* it deserves separate mention because of its already wide-

TABLE II

Some Fluorigenic Reagents of Value in Drug Analysis

Reagent		Derivatives Formed With:
Dansyl Chloride	5 dimethylamino-1 naphthalene sulphonyl chloride	primary, secondary amines phenols
NBD Chloride	4 chloro 7 nitro-benzofuran	primary and secondary amines
O Diacetyl-Benzene		primary amines
Fluram (Fluorescamine)	4 phenylspiro (furan -2(3H),1' -phthalan)-3,3'-dione	primary amines

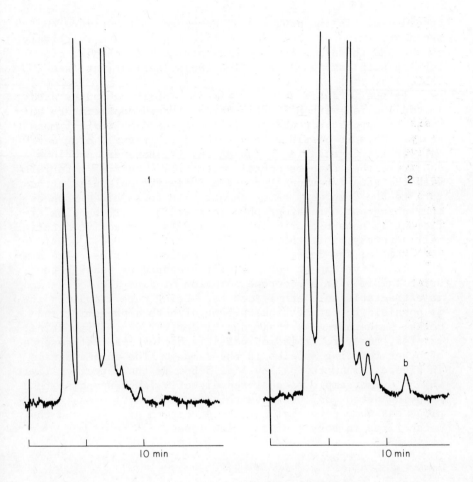

*Fig. 1. Liquid Chromatograms of Dansylated Urine Extracts.
1. Blank urine. 2. Urine spiked at the 5 ppm level (a = amphetamine, b = methyl amphetamine as their dansyl derivatives).*
Chromatographic Conditions.
Column: 25 cm x 4.9 mm id packed with Partisil 5 modified with octadecyltrichlorosilane.
Solvent: Methanol:water 85:15
Flow Rate: 1 ml/min at 1100 p.s.i.
Detector: Perkin Elmer 1000 fluorimeter with a "home-made" flow cell. λex 339 nm, λem 420 nm.

spread use. In the drug field, compounds with primary and secondary amino or phenolic functional groups are relatively common and there are several reagents available which can convert such compounds into fluorescent derivatives (see Table II).

In the forensic area fluorigenic labelling can be used in conjunction with HPLC to provide a rapid confirmatory analysis to support screening techniques based on radio-immuno assay. This is possible because most fluorescent drug derivatives are detectable at sub ng levels, however, problems can arise when samples contain naturally occurring materials with the same functional groups and similar molecular structure to the compounds being sought. In these circumstances the subsequent chromatographic separation needs to be as efficient as possible or high blank levels of reactive materials will severely limit the sensitivity of the method. Consider, for example, the detection of amphetamine in urine. The dansyl derivative can be very readily prepared by heating the urine sample with an acetone solution of dansyl chloride and is extractable into ethyl acetate. Using pure standards it is possible to detect about 0.5 ng of both amphetamine and methyl amphetamine as their dansyl derivatives. With urine samples, however, it is not possible because of the presence of other reacting species in the sample. The chromatograms in Figure 1 show that a level of 5 ppm is the detection limit when a urine sample is analysed directly (corresponding to 2 ng being detected). Despite this type of problem the technique has much potential for the detection of drugs and their metabolites in body fluids at low level.

Ion-Pair Fluorescence

For several years Scandinavian workers have been systematically studying the principles of ion-pair extraction, and a comprehensive review has recently been made of its application in the analysis of drugs (7). Chromatographers are becoming increasingly aware of the potential of the technique for enhancing chromatographic separations.

The basis of ion-pair extraction is that an organic ion Q^+ can be transferred from aqueous to organic solution by the addition of an ion with opposite charge X^-, and extraction of the ion-pair QX with a suitable organic solvent.

$$Q^+_{aq} + X^-_{aq} \rightarrow QX_{org}$$

The principle has been applied to the extraction of basic drugs from aqueous solution, and if a fluorescent counterion is produced from a compound such as anthracene 2 sulphonic acid (λex 305 nm λem 410 nm) the ion-pair will display similar fluorescence.

Our studies in this area are at an early stage but it is possible that ion-pairs produced from basic drugs could be separated chromatographically, and if the ion-pair displays fluorescence it could be detected to provide the basis of a sensitive analytical method.

REFERENCES

1. Wheals, B.B. (1976). Proceedings of the Analytical Division of the Chemical Society. To be published.
2. Jane, I. and Wheals, B.B. (1973). *J. Chromatog.* **84**, 181.
3. Jane, I. (1975). *J. Chromatog.* **111**, 227.
4. Christie, J., White, M. and Wiles, J. (1976). *J. Chromatog.* To be published.
5. Jane, I. and Taylor, J.F. (1975). *J. Chromatog.* **109**, 37.
6. Katz, S. and Pitt, W.W. (1972). *Anal Lett.* **5**, 177.
7. Schill, G. (1974). *In* "Ion Exchange and Solvent Extraction", Vol. 6, Chp. 1, (J.A. Marinsky and Y. Marcus, eds), Marcel Dekker.

Appendix

Type	Use	Functionality
Pellicular Silica (Active)	Normal Phase LSC and LLC	Silanol
Pellicular Silica (Inactive)	Normal Phase LLC	That of a Range of Stationary Phases
Microparticulate Silica	Normal Phase LSC and LLC	Silanol
Pellicular Alumina	Normal Phase LSC and LLC	Alumina
Microparticulate Alumina	Normal Phase LSC and LLC	Alumina
Pellicular Silica with Bonded Phase	Reverse or Normal Phase LSC	$-C_1, -C_{18}, -C_6H_5, -CN$, Polyethyleneoxide, Ether, Ethylnitrile
Porous Silica with Bonded Phase	Reverse or Normal Phase LSC	$-C_{18}, -C_6H_5, -CN$, Polyethyleneoxide
Microparticulate Silica with Bonded Phase	Reverse or Normal Phase LSC	$-C_1, -C_8, -C_{18}, -C_6H_5$, $-p$-allyl-phenyl, $-NO_2$ Alkylnitrile, Fluoroether, $-NMe_2$, Alkylamine
Glass Beads or Pellicular Silica with Bonded Anionic or Cationic Phase	Anion or Cation Exchange Chromatography	$-NH_2$ (Weak Base) $-NR_3^+$ (Strong Base) $-SO_3^-$ (Strong Acid)
Microparticulate Resin or Silica with Bonded Anionic or Cationic Phase	Anion or Cation Exchange Chromatography	$-NH_2, -NMe_2$ (Weak Base) $-NR_3^+, -NMe_3^+$ (Strong Base) Iminodiacetate (Weak Acid) $-SO_3^-$ (Strong Acid)
Porous Glass, Silica or Resin GPC Packing	Gel Permeation Chromatography	Molecular Exclusion

HPLC Packing Materials, their Properties and Use

Particle Size (μm) and Shape	Surface Area (m^2/g)	Polarity	Other
30-50 Irregular or Spherical	4-25	High	-
25-53 Irregular or Spherical	1-10	High or Medium	-
2-20 Irregular or Spherical	200-600	High	-
37-44 Spherical	4-8	High	-
5-20 Irregular or Spherical	70-95	High	-
25-53 Spherical	4-25	Low or Medium	-
37-75 Spherical	-	Low, Medium or High	-
5-13 Irregular or Spherical	200-500	Low, Medium or High	-
25-53	-	-	Ion Exchange Capacity (μequiv./g) 5-100
5-25	-	-	1-5
5-75	-	-	Pore Size 40-107 Å

A fuller Account of HPLC Columns and their Packings is given by Dr. R. E. Majors, International Laboratory 1975, Nov. Dec., p. 11.

INDEX

The Index is in three parts. The first lists compounds; no attempt has been made to index separately the various barbiturates, bile pigments, porphyrins, steroids etc. The second part lists the column packings which have been used and the third part is concerned with general matters, mainly techniques.

COMPOUNDS

Acetylcodeine, 205

Amino-acids, 24

Aminomethylpyrromethanes and derivatives, 64

Amphetamines, 216

Ascorbic acid, 205

Barbiturates, 144, 150, 158, 163, 166, 203

Benzodiazepines, 150, 171

Benzoic acid, 205

Bile pigments, 98

Bilirubin, 97, 169

Bitter principles, 46

Cannabis, 206

Carbamazepine, 144, 156, 171 metabolites, 160

Carboxylic acids, 46

Catecholamines, 120, 122, 127, 131, 138

Cephalosporins, 185, 191

Chlorophenols, 195

Chloroquinones, 195

Cholesterol and esters, 13, 34

Citric acid, 205

Corticosteroids, 59

Cytochrome c, oxidised, 21

Diphenylhydantoin and metabolite, 144, 145, 150, 158, 163

Dopa and derivatives, 127

Dopamine and derivatives, 121, 122, 127, 138

Epinine derivatives, 122

Fatty acids, 13

Glucose, 41

Glycosides, 46

Glutethimide, 150

Hydrochlorothiazide, 175

5-Hydroxytryptamine and derivatives, 121, 138

5-Hydroxytryptophane and derivatives, 121

Imidostilbene, 156
Indoles, 120
Isomaltose, 41
Lipids, 13, 34
Lipoproteins, 34
Lysergic acid diethylamide and derivatives (LSD), 205, 212
Melatonin and derivatives, 120
Methaqualone, 150
Morphine and derivatives, 213
Nicotine-adenine diphosphate, 113, 115
Nucleotides, 7, 111, 113, 115
Oestrogens, 51
Oligosaccharides, 42
Phenacetin, 144
Phenothiazines, 5
Phospholipids, 13, 34
Porphyrins, 66, 71, 81, 87
  copper complexes of, 76
Primidone, 144
Propoxyphene, 150
Proteins, 21
Saccharin, 205
Salicylates, 150, 205, 206
Serotonin, 121, 138
Steroids, 13, 34, 45, 51, 59
Succinimides, 144

Tetrahydrocannabinol, 206
Thebacon, 205
Tricyclic antidepressants, 6, 171
Triterpenes, 49
Tyramine and derivatives, 127, 137, 138, 139

COLUMN PACKINGS

ACA-22, 34
Acrylic-based ion exchangers, 21
Alumina, 88
Amberlite, LA-1, 54
Aminex-7, 133
Bondapak AX/Corasil, 123
Bondapak C_{18}/Corasil, 88, 120, 166
Bondapak Phenyl/Corasil, 123
µ Bondapak/Carbohydrate, 41
µ Bondapak C_{18}, 53, 60, 67, 83, 166, 175, 186
µ Bondapak NH_2, 51, 60
C3 ion exchange resin, 24
Cellulose-based ion exchangers, 21
Corasil II, 66, 71, 88, 98, 120, 166, 171
Diatomaceous earth, 54
Hydroxylapatite-based ion exchangers, 21
Lichrosorb RP8, 155
Merckosorb SI 60, 71

Micropak SLC-3, 132

Micropak SLC-11, 132

Micropak SI 10, 132

Micropak-CH, 132

Micropak-CN, 132

Octadecylsilyl silica, 126, 202, 205, 206, 209

ODS-Silex 1, 143

Ostion 0802 ion-exchange resin, 24

Ostion LGKS 0803 ion-exchange resin, 28

Partisil-10, 159

Partisil-5 ODS, 206, 215

Partisil-10 SCX, 133, 137, 203

Pellamidon, 88

AS Pellionex SAX, 110, 188

HS Pellionex SCX, 133

HC Pellionex SCX, 133

Polystyrene-divinylbenzene based ion exchangers, 21, 24

Porasil A-60, 73

Porasil C, 88

Porasil T, 90

μ-Porasil, 46, 52, 64, 66, 67, 81, 83, 99, 195

Rank-Hilger ion exchange resin, 24

Sepharose CL-4B, 34

Silica gel H, 33

Silicic acid, 13

NH_2-Silica, 8

SAS-Silica, 6

Spherisorb A-20Y, 5

Spherisorb ODS, 133, 139

Vydac TM, 123

Zipax, SAX, 188
 SCX, 188, 203

GENERAL

Adsorption systems, 5

Biological materials, direct analysis of, 128, 205, 206, 209

Colorimetric analysis, 13, 29

Column efficiency, 3, 188, 204

Column performance, 3

Competitive protein binding, 60

Detectors
 electrochemical, 1
 flame ionisation, 12, 34
 fluorimetric, 205, 206, 211
 moving wire, 11, 34
 photometer, fixed wavelength, 1, 47, 88, 120, 131, 206
 photometer, variable wavelength, 1, 5, 51, 60, 64, 71, 81, 88, 99, 110, 125, 143, 155, 165, 175, 186, 198

 refractive index, 41, 47, 99

Detergents, in solvent systems, 6, 126

Equipment, problems of, 1

Flow programming, 34, 81, 99

Gel permeation systems, 34

Injection, problems of, 2

Ion exchange systems, 6, 21, 24

Ion pair chromatography, 6

Mass spectrometry, 73, 205, 209

Preparative separations, 19, 41, 49, 73

Partitioning systems, 4

Radioimmunoassay, 119, 209

Radioisotope tracers, 61, 110, 115, 122, 134, 137, 209

Scintillation counting, 209

Solvent programming, 18, 24, 72, 90, 110, 132
  stepwise, 13, 24, 120

Theoretical plate height, absolute, 3, 188

Theoretical plate height, reduced, 3